Machinery Component Maintenance and Repair

Machinery Component Maintenance and Repair

Editor

Chinmay Tripathi

Machinery Component Maintenance and Repair
Edited by **Chinmay Tripathi**

Printed in 2017

ISBN: 978-1-68117-408-2

Library of Congress Control Number: 2015941614

© 2016 by
SCITUS Academics LLC,
616, Corporate Way, Suite 2, 4766,
Valley Cottage, NY 10989

www.scitusacademics.com

This book contains information obtained from highly regarded resources. Copyright for individual articles remains with the authors as indicated. All chapters are distributed under the terms of the Creative Commons Attribution License, which permits unrestricted use, distribution, and reproduction in any medium, provided the original author and source are credited.

Notice

Reasonable efforts have been made to publish reliable data and views articulated in the chapters are those of the individual contributors, and not necessarily those of the editors or publishers. Editors or publishers are not responsible for the accuracy of the information in the published chapters or consequences of their use. The publisher believes no responsibility for any damage or grievance to the persons or property arising out of the use of any materials, instructions, methods or thoughts in the book. The editors and the publisher have attempted to trace the copyright holders of all material reproduced in this publication and apologize to copyright holders if permission has not been obtained. If any copyright holder has not been acknowledged, please write to us so we may rectify.

Contents

Preface vii

Chapter 1   Failure Analysis of Machinery Component by Considering External Factors and Multiple Failure Modes – A Case Study in the Processing Industry 1
Rosmaini Ahmad, Shahrul Kamaruddin, Ishak Abdul Azid, and Indra Putra Almanar

Chapter 2   The MRN Complex in Double-strand Break Repair and Telomere Maintenance 31
Brandon J. Lamarche, Nicole I. Orazio, and Matthew D. Weitzman

Chapter 3   Standardized Cost Estimation for New Technologies (SCENT) - methodology and Tool 83
Stanil Y. Ereev and Martin K. Patel

Chapter 4   Removal of Greases and Lubricating Oils from Metal Parts of Machinery Processes by Subcritical Water Treatment 121
Walter J. Weber Jr and Han S. Kim

Chapter 5   The Intricate Structural Chemistry of Base Excision Repair Machinery: Implications for DNA Damage Recognition, Removal, and Repair 145
Kenichi Hitomi, Shigenori Iwai, and John A. Tainer

Chapter 6   An Approach to Performance Assessment and Fault Diagnosis for Rotating Machinery Equipment 199
Xiaochuang Tao, Chen Lu, Chuan Lu, and Zili Wang

Chapter 7	**A Framework to Determine the Effectiveness of Maintenance Strategies Lean Thinking Approach** 237	
	Alireza Irajpour, Ali Fallahian-Najafabadi, Mohammad Ali Mahbod, and Mohammad Karimi	

Citations... 267

Index... 271

Preface

Maintenance and repair of machinery in a petrochemical process plant was defined in a preceding volume as simply "defending machinery equipment against deterioration." Machinery maintenance can often be quite costly in a petrochemical plant operation. Prior to the publication of the first two volumes of this series, very few studies were available describing quantitative or objective methods for arriving at the optimization of the four strategies. Though our readers should not expect detailed contributions to those subjects in this volume, we did opt to include an overview section describing the maintenance philosophy practiced in a large multi-plant corporation which makes effective use of centralized staff and computerized planning and tracking methods. The term repair are often reserved for maintenance actions that improve the conditions of an item, but may or may not establish "good as new" condition. In fact, repair is strictly reserved for maintenance of an item that has reached a defined failed state or defect limit.

Editor

Failure Analysis of Machinery Component by Considering External Factors and Multiple Failure Modes – A Case Study in the Processing Industry

Rosmaini Ahmad[1], Shahrul Kamaruddin, Ishak Abdul Azid[2], and Indra Putra Almanar[3]

School of Mechanical Engineering, Universiti Sains Malaysia, Engineering Campus, Nibong Tebal, 14300 Penang, Malaysia

ABSTRACT

This paper proposes a failure analysis method by integrating the Failure Mode Effect and Criticality Analysis (FMECA) and Failure

Time Modelling (FTM) based on Proportional Hazard Model (PHM). The objectives of FMECA application are twofold: to classify the censored and uncensored data based on the criticality measure of FMECA, and to identify possible external factors (covariate effects) based on cause and effect assessments of FMECA. FTM based on PHM is applied to analyze statistically the censored and uncensored failure time data by considering the effects of external factors. The applicability of proposed methodology is illustrated in a case study in the processing industry. The analysis results of FTM based on PHM by considering external factors present a more reliable hazard rate and lifetime or mean-time-to-failure of the targeted component, hence enabling engineers to plan a more effective maintenance strategy.

INTRODUCTION

Much has been written in the literature about failures and their classifications. Their clarifications are based on various perspectives. For example, Nowlan and Heap [1] divided failure into two types: functional and potential. Functional failure is the inability of an item (or the component containing it) to meet a specified performance standard, whereas potential failure is an identifiable physical condition indicating an imminent functional failure. Badía and Berrade [2] classified machine failure into two types, namely, minor, and catastrophic. Minor failures can be avoided with minimal repair, while catastrophic failures are removed by a major or perfect repair. According to SS-EN 13306 [3], failure is the termination of the ability of an item (i.e., any part, component, device, subsystem, functional unit, component, or system that can be individually considered) to perform a required function. In other words, failure can be referred to as an event or process of component deterioration.

The natural process of a mechanical component/system failure relates to its age, specifically the wear and tear effects. In many cases, increasing in aging is followed by reduced component performance (degradation) [4]. According to Blischke and Murthy

[5], the consequences of failure are many and varied, but most failures create economic impact. In manufacturing industries, production machine failures can create many inconveniences. It may be responsible for machine downtime, low availability, customer dissatisfaction, increased maintenance time and production costs, lower product quality, and delivery time delay [6]. In chemical or nuclear plants, failures might be very expensive (catastrophic effects) and in some cases, are not allowed at all due to safety issues. Although, failure cannot be totally avoided, the related risks and effects can be minimized and controlled by implementing effective and efficient maintenance practice.

The decision-making process towards the improvement of a component/system (either to redesign or revised maintenance policy) must start with failure analysis. For the component/system in the operating stage, failure analysis based on their field failure data is useful in identifying, classifying, and calculating the criticality of physical and functional failures. For these purposes, the application of semi-qualitative failure analysis tool, failure mode and effect analysis (FMEA) and/or Failure Mode Effect and Criticality Analysis (FMECA) are helpful.

Although, FMEA/FMECA is typically used as a problem prevention tool to improve the reliability of a system/component, especially at the design stage [7], it is also useful in the operating stage [8]. The FMEA/FMECA attempts to evaluate comprehensively the effects of each failure mode of every component on a system. In addition, FMEA/FMECA becomes more effective if it is carried out based on "brainstorming" approaches among expert personnel. Therefore, the entire physical characteristics of failure can be identified, classified, and clarified properly. Ebeling [9] presents the standard process in performing the FMECA, which starts with component/system definition; failure mode, failure cause and failure effect identification; classification of severity; estimation of the probability of occurrence; computation of criticality index of each failure mode; and determination of corrective action. The other sources of standard process in performing the FMEA/FMECA are given in [10] and [11].

The literature presents the related research of FMEA/FMECA, which includes their applications and extended version. For example, Becker and Flick [12] presented an approach of FMECA as a distributed computing system for air traffic control. Teng and Ho [13] developed an approach to integrate FMEA, product design, and process control to one complete closed loop to establish an overall quality control plan. A machining operation in a hybrid inflator production process has been used as an example to show the benefit of the developed approach. Xu et al. [14] presented a fuzzy-logic-based method for better application of FMEA, in which they used a turbocharger system of a diesel engine to illustrate the feasibility of the proposed method. Puente et al. [15] described an alternative way of applying FMEA to a wide variety of problems; they then proposed a fuzzy decision model to improve the traditional FMEA in calculating the risk priority number (RPN). Yang et al. [16] also improved the FMEA model by focusing on developing a systematic evaluation and improvement mechanism to locate the RPN of the items/products/components for semiconductor related industries in Taiwan. Teng et al. [17] studied the application of FMEA in a collaborative environment by focusing on the implementation process. Their study offered guidelines for better FMEA applications for the manufacturing industry, such that the companies can adopt their FMEA process into a collaborative supply chain environment. Arabian-Hoseynabadi et al. [18] applied the FMEA in studying the reliability of many different power generation systems. They compared the quantitative results of an FMEA and reliability field data from real wind turbine systems and their assemblies, and the results helped establish relationships that are useful for the design of wind turbines. Xiao et al. [19] studied the application of FMEA for the case of multiple failure modes analysis by focusing on RPN calculation. Their study extended the work carried out by Pickard et al. [20], who revised the calculation of RPN by multiplying it with a weight parameter, which characterized the importance of the failure causes within the system. The effectiveness of the method is demonstrated with numerical examples. Failure Time Modelling (FTM) is another example of failure analysis that can be carried out using field failure data. FTM is also known as a reliability analysis that statistically analyzes the failure characteristics of the component based on time-to-failure (TTF) data through particular failure distribution based on the probability approach. In FTM, the characteristics of the failure time is represented by the random variable "T" (time-to-failure) and can be formed into different failure time characteristics functions, such as hazard rate, reliability, cumulative distribution, and probability density [9]. The hazard rate function presents the characteristics of failure distribution, whether in increasing, constant, or decreasing failure rates. The reliability function shows the reliability trend (failure distribution) of the component by given time, while the cumulative distribution function presents the failure distribution of the component in cumulative form. Finally, the probability density function illustrates the shape of failure distribution. Each of these failure functions can be modelled by different forms based on different failure times or distribution models.

In the literature, FTM has been carried out to analyze the failure characteristics for a variety of cases. For example, Ferreira et al. [21] performed the FTM of roller bearings used in railway ore transportation wagons by employing the Weibull distribution model. In that study, they divided data from 47,000 failed bearings into seven groups and then used these to determine failure distribution. They also used two methodologies in this FTM to assess the bearing nominal lifetime: method I standard procedure (proposed by the bearing manufacturer) and method II (developed by the authors). By comparing the results, Ferreira et al. found that method II is more flexible because it allows the control of the parameters directly involved in the bearing failure phenomenon. Liberopoulos and Tsarouhas [22] carried out FTM to investigate the failure characteristics of an automated pizza production line; they applied the Weibull distribution model to compute related parameters, such as failure rate and found that most failures have a decreasing failure rate, while a few have an almost constant failure rate. Their study results can guide manufacturers of food product machinery as well as bread and bakery products to improve the design and operation of their respective production lines.

Tsarouhas et al. [23] analyzed the failure characteristics of a strudel production line. For 16 months, they collected data from the production line and used these to calculate the reliability and hazard rate modes for all workstations and the entire production line. They claimed that the study is useful in assessing the current conditions and predicting reliability for improving the production line maintenance policy. In another paper, Tsarouhas et al. [24] performed the FTM for cheese (feta) production line in a Greek medium-size company.

Similar to previous studies, they also calculated the failure characteristics towards reliability and hazard rate modes at the entire production line. The results from their study are useful tools in assessing the current conditions and predicting the reliability for upgrading the maintenance strategies of the production line. In a recent paper, Tsarouhas and Arvanitoyannis [25] presented the results of FTM carried out on a peach production line. The results of FTM (pertaining to the reliability and hazard rates) for the case study suggest that the current maintenance policy is not adequate for the entire peach production line and more effort is required for improving the operation management of the line.

In FTM, the classification of censored and uncensored data and the assumption of TTF data play critical roles in obtaining more reliable results. According to Ebeling [9], censored data refer to the data that are incomplete because a component is removed from consideration prior to their failure. For example, a component may fail because of other failure modes than the one being measured. Thus, the proper classification of censored and uncensored data is very important in ensuring that the FTM achieves more reliable results. The classification of the censored and uncensored data and the identification the most possible external factors in the proper manner is significant for FTM, but this aspect is rarely discussed in the literature. In other words, a tool is required for the classification and identification processes, and FMECA can possibly be applied in such situations. Meanwhile, the assumption of failure time data used in FTM is another issue that can affect the failure analysis results. In most cases, failure time data are assumed as being dependent on time or age. However, in reality, the failure of any component is also influenced by external factors, such as human errors and environmental factors. For instance, Oyebisi [26] studied the effects of high temperature and humidity on electronic component and found that drastic changes (increase or decrease) in the working temperature and humidity affect the performance of the component due to failure.

Martorell et al. [27] studied the effects of the environment on the lifetime of the mechanical components of Nuclear Power Plant (NPP) and reported that an overdose of radiation from the nuclear process has significant effects on the material properties of mechanical component, thus reducing the component lifetime. Dhillon and Liu [28] stated that component failure may occur due to maintenance error by carrying out incorrect repair or preventive action, such as incorrect calibration and repair procedure. Meanwhile, Latorella and Prabhu [29] presented the most common maintenance errors related to industry, including incorrect installation of components, fitting of the wrong parts, electrical wiring discrepancies (including cross connection), loose objects (tools), and inadequate lubricant. Thus, considering the effects of external factors in FTM is important in determining more reliable results of failure analysis.

The current paper deals with failure analysis of FTM by considering the external factors as well as the censored and uncensored data. The methodology of failure analysis by integrating the FMECA and FTM based on Proportional Hazard Model (PHM) is proposed. FMECA is applied for censored and uncensored data classification and external factor identification, while PHM is incorporated in FTM to describe the relationship of TTF data and identify the external factors. This paper is organized as follows. Section 2 describes the proposed failure analysis methodology. Section 3 presents the usefulness of the methodology based on a case study at a processing industry. Finally, Section 4 concludes the failure analysis results and suggests future research directions.

PROPOSED FAILURE ANALYSIS METHODOLOGY

Fig. 1 shows the structure of the proposed failure analysis methodology. The basic idea of the methodology integrates the FMECA and FTM based on PHM. In this proposed methodology, the objectives of FMECA are twofold: to classify the censored and uncensored data, and identify the possible external factors (i.e., covariate effects). The objective of FTM based on PHM is to analyze statistically the failure time (time-to-failure) data based on censored and uncensored classification and also consider the effects of external factors. The detail application of FMECA and FTM based on PHM is given in the following sections.

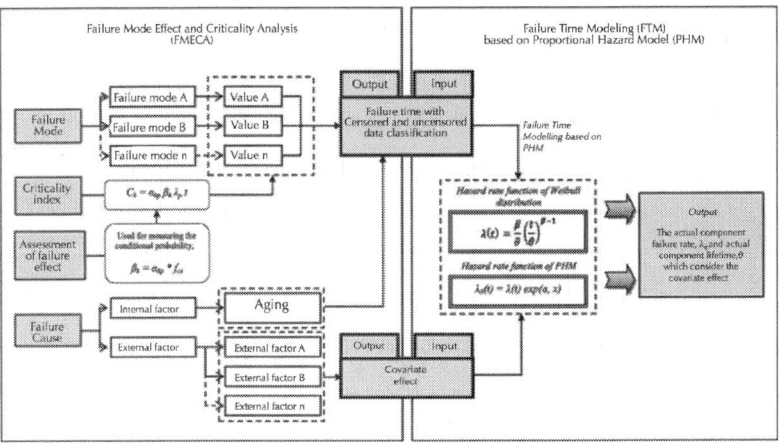

Figure 1: Structure of the proposed failure analysis methodology – integration of FMECA and FTM based on PHM.

FMECA

Referring to Fig. 1, the application of FMECA in the proposed methodology focuses on four related steps towards achieving its specific objectives. The four steps in the analysis of FMECA are failure mode identification, failure cause identification, assessment of failure effect, and criticality index calculation of each failure mode.

The first analysis step of FMECA applied in the proposed methodology is failure mode identification, which can be classified as a qualitative type analysis. Failure modes of a failed system/component can be identified and classified either based on the physical (e.g., fractures, break, loss of contact, etc.) or functional (e.g., power loss, being out of tolerance, etc.) points of view [9]. In the proposed methodology, failure mode identification is an important step for classifying the censored and uncensored data to be further used in FTM based on PHM.

The second analysis step in the FMECA application is the failure cause identification, which is classified as a qualitative type analysis. Generally, the failure cause identification step aims to identify the possible cause or causes that contribute to each failure mode. In this proposed methodology, the failure cause identification of each failure mode can be grouped into internal and external factors. The internal factor refers to the natural process of component failure, also known as aging factor brought about by certain causes. For example, the aging factor for most rotating components (e.g., bearing, gear, belt, etc.) is brought about by the effects of wear and tear. Meanwhile, the external factor refers to the factors (e.g., frequency of maintenance activities, type of products, etc.) that contribute to the failure process of a component. In other words, external factors can make the failure process slower or faster than the estimate of the original equipment manufacturer (OEM). Thus, the age of each component lifecycle may significantly vary due to the effect of external factors during operating time. In the proposed methodology, the possible external factors (also known as covariate effects) that have been identified are analyzed further in FTM based on PHM in order to estimate its effects to the component failure.

The third analysis step of FMECA applied in the proposed methodology is the assessment of failure effect. Similar to previous analysis steps, the assessment of failure effect is classified as a qualitative type analysis. This assessment verbally analyzes the effect of each failure mode on the system (entire machining process) and classifies it into several categories, also known as the classification of severity categories. Ebeling [9] defined the classification of severity categories below.

Category I:	Catastrophic – significant system failure occurs that can result in injury, loss of life, or major damage
Category II:	Critical – complete loss of system occurs, performance is unacceptable
Category III:	Marginal – system is degraded, with partial loss in performance
Category IV:	Negligible – minor failure occurs, with no effect on acceptable system performance

The final analysis step of FMECA application is the criticality index estimation for each failure mode that has been identified. Criticality index estimation is classified as a quantitative type that combines the probability of the failure mode occurrence with its severity ranking. The criticality index of each failure mode can be calculated using the formula [9]:

$$C_k = \alpha_{kp} \beta_k \lambda_p t \qquad (1)$$

where C_k is the criticality index for failure mode k; α_{kp} the fraction of the component p's failure having failure mode k (i.e., the conditional probability of failure mode k given that component p has failed); β_k the conditional probability that failure mode k results in the identified failure effect; λ_p the failure rate of component p; and t is the duration of time used in the analysis.

According to Eq. (1), the parameter, S, of α_{kp}, λ_p, and β_k can be respectively estimated as follows:

$$\alpha_{kp} = \frac{f_{kp}}{f_{total-p}}, \qquad (2)$$

$$\lambda_p = \frac{f_{total-p}}{t}, \qquad (3)$$

$$\beta_k = \alpha_{kp} * f_{cs}, \qquad (4)$$

where f_{kp} is the number of failure of component p having failure mode k; $f_{total-p}$ the total number of failures used in the analysis for the component p; and f_{cs} is the probability value of severity classification given in Table 1.

Table 1: Probability value of severity classification

Category of severity classification	Probability value, f_{cs}
Category I: Catastrophic – significant system failure occurs that can result in injury, loss of life, or major damage	1.0
Category II: Critical – complete loss of system occurs, performance is unacceptable	0.75
Category III: Marginal – system is degraded, with partial loss in performance	0.50
Category IV: Negligible – minor failure occurs, with no effect on acceptable system performance	0.25

From the value of criticality index, C_k of each failure mode is then compared with each other in the process of classifying censored and uncensored data. The classification of the data is then analyzed further in FTM based on PHM.

FTM Based on PHM

FTM is a process of failure time analysis to identify the characteristics of component failure (time-based) based on a particular failure distribution model [30]. The characteristics of a component failure can be represented by the random variable, T (time to failure), which can be used to represent different failure time functions, such as reliability, probability density, and failure rate or hazard rate functions. These failure functions can be presented in different forms based on different failure distribution models, and the most popular distribution model applied is the Weibull distribution model. According to Ghodrati and Kumar [31], the Weibull distribution model is a flexible model (can model any state of component lifetime, whether decreasing, constant, or increasing failure rate) that can be applied for characterizing most of the mechanical components. Eqs. (5), (6) and (7) present the reliability, probability density, and hazard rate functions based on the Weibull distribution model, respectively, as [9]:

$$R(t) = e^{-(t/\theta)^\beta}, \tag{5}$$

$$f(t) = \frac{\beta}{\theta}\left(\frac{t}{\theta}\right)^{\beta-1} e^{-(t/\theta)^\beta}, \tag{6}$$

$$\lambda(t) = \frac{\beta}{\theta}\left(\frac{t}{\theta}\right)^{\beta-1}, \tag{7}$$

where $R(t)$ is the reliability function; $f(t)$ the probability density function; $\lambda(t)$ the failure rate or hazard rate function; e the exponent; β the shape parameter of the Weibull distribution model; ϑ the scale parameter of the Weibull distribution model; and t is the failure time.

The integration of FTM and PHM is applied in the proposed methodology to analyze/model the failure time (time-to-failure) data by considering the effects of external factors (covariates). PHM is a regression type model describing the relationship between the failure data based on an internal factor (time-dependent) and the change in the effect of the covariates (external factor) [32]. The basic assumption of PHM is that the actual hazard rate, $\lambda_0(t)$, of a component is the product of a baseline (time-dependent) hazard rate $\lambda(t)$ and a positive (time-independent) functional term exp(a, x), which is basically independent of time, and incorporates the effects of a number of covariates [33]. Mathematically, the failure rate function of PHM can be expressed as:

$$\lambda_0(t) = \lambda(t) \exp(a, x) \tag{8}$$

where a is a row vector consisting of the covariates, and x is a column vector consisting of the regression parameters. The actual hazard rate, $\lambda_0(t)$ (observed hazard rate) changes when the covariates change. Referring to Eq. (8), the value of the covariate parameter, x, can be determined based on the likelihood function given by Kalbfleisch and Prentice [34] as:

$$L(x) = \prod_{j=1}^{k} L_j(x) = \prod_{j=1}^{k} \frac{\exp(S_j x)}{\left[\sum_{m \in F(t_j)} \exp(a_m x)\right]^{d_j}} ; \qquad (9)$$

where $L(x)$ is the conditional probability that the failure occurred at the time t_j; x the covariate parameter; k the number of time-to-failure (TTF); S_j the sum of the covariate observed at the failure time t_j; m the number of items in the risk set at time t_j; a the row vector of covariate; d_j the number of tied failures; and $F(t_j)$ is the risk set of items that function just prior to the observed failure at time t_j.

The value of covariate parameter, x, that maximizes Eq. (9) can be obtained using numerical methods. The estimated value of the covariate parameter is then tested for its significance with 95% (0.05) confidence intervals in order to verify whether or not the covariate has any significant effect on the component failure.

The concept of covariates that have an effect on baseline (time-dependent) hazard rate (t) is shown in Fig. 2. Baseline hazard rate is the failure rate without considering the covariate effect (only internal factor effects), whereas the observed (actual) hazard rate, $\lambda_0(t)$, is the failure rate influenced by considering the covariate effect. Referring to Fig. 2, the effects of covariates may increase or decrease the observed hazard rate. For example, the environmental effects, such as dust, heat and radiation, may increase the component hazard rate.

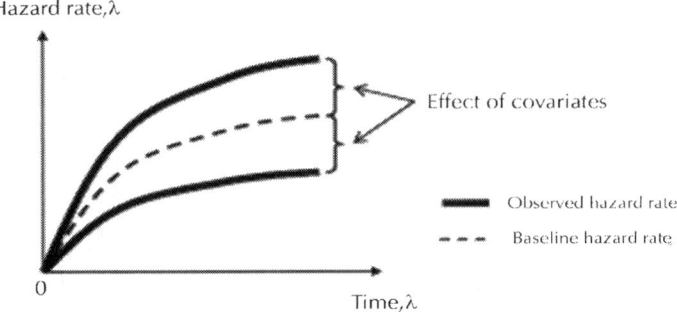

Figure 2: Effect of covariates on the hazard rate of the component.

Based on the proposed methodology structure (Fig. 1), the baseline hazard rate, λ(t), is the function of FTM, while exp(a, x) is the function of covariate modelling. The baseline hazard rate function, λ(t), can be modelled based on the Weibull distribution model, which is widely used in conventional FTM (without considering the covariate effect). Therefore, Eq. (10) presents the application of the Weibull distribution model and PHM as:

$$\lambda_0(t) = \frac{\beta}{\theta}\left(\frac{t}{\theta}\right)^{\beta-1} \exp\left(\sum_{j=1}^{n} a_j x_j\right), \qquad (10)$$

where $\lambda_0(t)$ is the actual hazard rate function by considering the covariate effect; β the shape parameter of the Weibull distribution model; ϑ the scale parameter of the Weibull distribution model; t the failure time; a_j the covariate, j; and x_j the covariate parameter, j.

The output of the FTM based on PHM is the value of covariate parameters affecting the baseline hazard rate function, λ(t). It is important to point out that the value of covariate parameters change the value of scale parameter, θ, of the Weibull distribution model, which refers to the mean-time-to-failure (MTTF) of the component [31]. From the actual hazard rate function given in Eq. (10), the actual value of scale parameter, θ_0, by considering the covariate effects, can be obtained based on Eq. (11) as shown below [31]:

$$\vartheta_0 = \vartheta \left(\exp\left(\sum_{j=1}^{n} a_j x_j \right) \right)^{-\left(\frac{1}{\beta}\right)}, \tag{11}$$

where β is the shape parameter hazard rate; ϑ the scale parameter of baseline hazard rate; ϑ_0 the scale parameter of hazard rate by considering the covariate effects; a_j the covariate, j; and x_j is the covariate parameter, j.

The following sections illustrate the applicability of the proposed methodology in an industrial case study.

METHODOLOGY VALIDATION

The applicability of the failure analysis methodology presented in Section 2 is validated in an industrial case study at a processing industry in Penang, Malaysia. The company is classified under the wood-based type industry that produces various types of products for daily use. In the case study company, the production floor is the main area where major machining processes (e.g., cutting, embossing, rewinding, etc.) are carried out by production machines. The case study focuses on one of the machinery components of the cutting process system, which is the transmission belt. Fig. 3 shows the illustration of the transmission belt function in the cutting process system. The basic function of the transmission belt is to facilitate the high speed rotation of the cutting knife; it moves in an up and down direction to complete the cutting function. The function of the transmission belt is supported by two bearing pulleys, and the movement is powered by an electric motor.

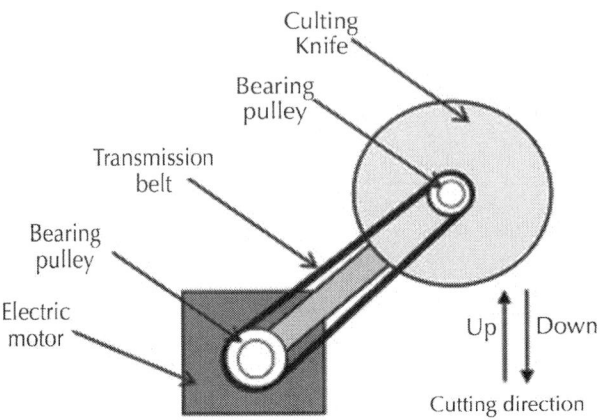

Figure 3: Illustration of the transmission belt function in the cutting process system.

The continuous mechanical stress due to the wear and tear effects of the transmission belt during operating time degrades the function of this component, thus requiring replacement. Based on the maintenance record, failure due to the unplanned replacement of the transmission belt disturbed the production schedule and increased the production costs significantly. Thus, designing a preventive maintenance (PM) policy by determining the right time to perform a replacement is the best solution to minimize the effects of any unplanned replacement of the transmission belt. To estimate the right PM time for transmission belt, failure analysis towards identifying its actual failure characteristics is needed. The following sections present and discuss the results of failure analysis using the proposed method.

Data Required

The data required for the proposed methodology validation are classified into two types: quantitative and qualitative. Quantitative data refer to failure times or time-to-failures (TTFs) of the transmission belt, where it is measured in the scale of working days. Meanwhile, qualitative data refer to the observed information, such as physical failure modes, possible failure causes (external factors), and failure effects. The data of TTFs and physical failure modes are collected based on maintenance records, while data related to the possible failure causes and effects are gathered via brainstorming sessions with the engineers and technicians from the maintenance and production departments. The initial data of TTFs and failure modes of transmission belt are presented in Table 2.

Table 2: Initial data of FMECA for the transmission belt

No. of failure	Quantitative	Qualitative
	Time between failure (working days)	Failure mode
1	28	Loss of grip
2	52	Loss of grip
3	42	Breakage
4	8	Loss of grip
5	14	Loss of grip
6	13	Loss of grip
7	47	Loss of grip
8	38	Loss of grip
9	25	Loss of grip
10	12	Loss of grip
11	50	Breakage
12	42	Loss of grip

RESULTS AND DISCUSSION

Table 3 summarizes the overall results of FMECA. Based on the number of failure records within a period of 371 working days, the failures of transmission belt are due to two types of failure modes, namely, loss of grip and breakage. From the brainstorming activity, the possible failure causes (external factors) that contribute to these failure modes may be influenced by dust effects, processing of different types of products, and pulley tension effects (unstable bearing). The reasons of selecting these three external factors can be explained below.

- *Dust effects* – Dust is one of the largest waste products exposed to the machine components (including transmission belt) during the cutting process. The transmission belt operates in open condition and with high speed, the transmission belt and pulley surfaces are directly exposed to dust that may reduce the grip between their surfaces. Therefore, dust effect is considered as the main cause of failure modes (i.e., "loss of grip" and "breakage").
- *Different type of product effects* – The main function of the transmission belt is to move (roll) the cutting knife to facilitate the cutting process (refer to Fig. 1). The products cut by the cutting knife are classified into two: hard and soft types. Logically, the cutting of a hard type product needs more force, thus it contributes to the transmission belt failure due to failure mode (i.e., "loss of grip" and "breakage").
- *Pulley tension (unstable bearing) effects* – This effect refers to the unstable bearing on the pulley that supports the function of the transmission belt. The failure record shows that even when the failed transmission belt is replaced, the bearing on the pulley is not replaced at the same time. The different replacement times affect the function of the transmission belt because the unstable bearing on the pulley generates inconsistent strains on the belt, thus resulting in transmission belt failure due to failure mode (i.e., "loss of grip" and "breakage").

Table 3: FMECA result

Critical component	Failure mode	Failure cause		Assessment of failure effect		Parameters of criticality index				Criticality index, C_k
		Internal factor	External factor	Failure effect	Classification of severity, f_{cs}	λ_p	β_k	α_{kp}	t(working day)	
Transmission belt	Loss of grip	Wear and tear	Dust	The pulley of the cutting knife cannot move properly, thus degrading the cutting process (leading to product quality problems)	III (0.50)	$=3.2 \times 10^{-2}$	0.42	0.83	371	4.14
			Product types (hard and soft)							
			Pulley tension (unstable bearing)							

Breakage	Wear and tear	Dust	When the transmission belt breaks, the cutting process completely stops	II (0.75)	0.13	0.17			0.26
		Product types (hard and soft)							
		Pulley tension (unstable bearing)							

For the assessment of failure effect, which is the next analysis step of FMECA, the brainstorming team finalized that the effects of failure modes, "loss of grip" and "breakage" actually vary. Using the severity classification of the failure effect assessment given in Table 1, the failure effect of failure mode "loss of grip" is categorized as marginal level, where the probability value of this category is 0.5. Meanwhile, the failure mode "breakage" is categorized as critical level, where the probability value of this category is 0.75.

The final analysis step of FMECA is the calculation of the criticality index, C_k, for each failure mode. From Table 3, by applying Eqs. (1), (2), (3) and (4), the criticality index for failure mode "loss of grip" is 4.14, while that for "breakage" is 0.26. The calculated criticality indexes indicate that the failure of the transmission belt due to "loss of grip" is significantly higher compared with "breakage." This result suggests that the next failure analysis (FTM based on PHM) must classify the TTFs due to "breakage" as censored data (incomplete data coded as 0) and the TTFs due to "loss of grip" as uncensored data (complete data coded as 1).

Table 4 presents the data used in FTM based on PHM, which are classified into three categories. The first category comprises the TTFs of the transmission belt. The second category is the classification of the censored and uncensored data of the TTFs. The third category refers to the external factor (covariate) parameters that have been identified in FMECA. According to the FMECA result presented in Table 3, dust effect, different product types, and unstable bearing pulley are the possible external factors (covariates) that contribute to the transmission failure. To consider these covariate parameters in FTM based on PHM, the factors are classified and codified in a special manner as follows.

- *Dust (DUST)* – If the component is exposed to extreme dust condition without (or less) any preventive action (cleaning) during the component lifecycle, the component is codified as "−1" (bad condition). If the component is recorded for cleaning activity extensively during the component lifecycle, it is codified as "+1" (good condition).

- *Related component factor (RCOMP)* – If the related component (e.g., the bearing in a pulley for supporting the belt operation) is replaced together with the critical component (refers to the case of transmission belt), it is codified as "+1" (good condition) and "−1" (bad condition).
- *Product type (PRODT)* – This covariate is formulated in a continuous form and is based on the product types (i.e., "hard and soft"). For example, assuming that the failure time of the critical component is 10 (in working shifts), then the type of product produced at each shifts (failure time = 10 working shifts) is recorded. For example, three shifts have produced soft types of product and seven shifts have produced hard types of product. Therefore, this covariate (PRODT) is codified as 0.7 (percentage), referring to a hard type product. In this case, the critical component needs more force (more risk to fail) to process the hard type product compared with the soft type product.

Table 4: Data classification for FTM based on PHM

No. of failures	Time between failure (working days)	Censored and uncensored data	Covariate		
			RCOMP	DUST	PRODT
1	28	1	−1	−1	0.6
2	52	1	1	1	0.4
3	42	0	−1	1	0.4
4	8	1	−1	−1	0.3
5	14	1	−1	−1	0.6
6	13	1	−1	−1	0.6
7	47	1	1	1	0.5
8	38	1	1	−1	0.6
9	25	1	1	−1	0.4
10	12	1	−1	−1	0.8
11	50	0	1	−1	0.6
12	42	1	−1	1	0.6
Total			−2	−4	6.4
Average			−0.167	−0.333	−0.533[a]

ᵃ[– < 0.5 < +].

Table 5 presents the result of covariate estimations for transmission belt failure analysis using Eq. (9). The covariates RCOMP and DUST are found to have a significant effect on the transmission belt failure based on the p-value less than 0.05 (95% confident interval), while the covariate PRODT does not have a significant effect because the p-value is larger than 0.05. The parameters of covariates RCOMP and DUST are –0.943 and –1.032, respectively.

Table 5: Estimation of covariates of transmission belt failure using PHM

Parameter	Estimate	SE	t-Ratio	p-Value
Final model summary				
RCOMP	–0.943	0.472	–1.998	0.046
DUST	–1.032	0.490	–2.104	0.035
	Estimate	Lower	Upper	
95.0% Confidence intervals				
RCOMP	–0.943	–1.868	–0.018	
DUST	–1.032	–1.993	–0.071	

After estimating the effects of external factors (covariates) on the failure times of the transmission belt (as presented in Table 5), the FTM—without considering the external factor—is carried out for the purpose of comparison. The main objective is to identify the values of scale parameter (lifetime) of the Weibull distribution model, where, as mentioned in Section 2.2, the effects of covariates can change the value of the scale parameter. Thus, the value of scale parameter, θ, without considering the covariates, is first calculated before the actual value of scale parameter, θ_0, is determined using Eq. (11).

This process is carried out using conventional FTM based on the parametric approach, and the result is presented in Table 6. The estimated shape parameter, β, is 2.1, while the value of scale parameter, θ, without considering the covariate effect, is 35.0. Using Eq. (11) and taking into account the estimated covariates as presented in Table 5, the actual value of scale parameter, θ_0, is calculated to be 27.6. This value refers to the MTTF or lifetime of the transmission belt by considering the covariate effects.

Table 6: Comparison of the Weibull distribution parameters with and without considering the covariate effects

Weibull distribution parameters	Without considering covariate effects, ϑ	With considering covariates effect, ϑ_0
Scale parameter,	35.0 (working days)	27.6 (working days)
Shape parameter, β	2.1	2.1

In addition, the results of FTM can be presented in other forms of failure characteristic functions, such as hazard rate, probability density, and reliability functions. In this case, it is interesting to show the effects of each of these failure functions with and without considering the covariate effects. Fig. 4, Fig. 5 and Fig. 6 present the trend of hazard rate, probability density and reliability functions, respectively. In these figures, the solid line represents the trends without considering the covariate effects, while the dotted line shows the trends with considering the covariate effects.

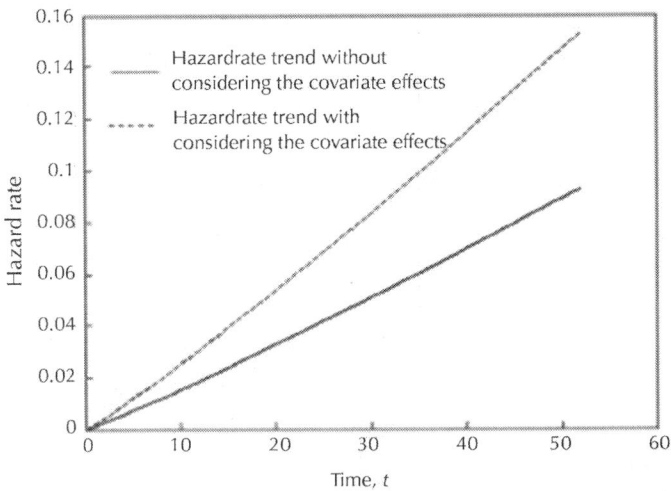

Figure 4: Effect of covariates on the Weibull hazard rate function.

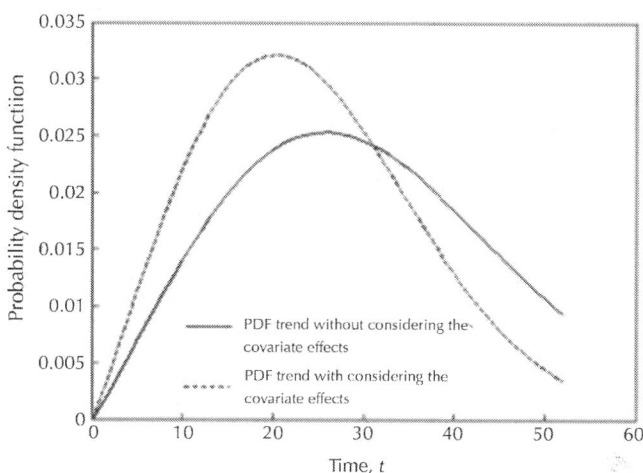

Figure 5: Effect of covariate on the Weibull PDF.

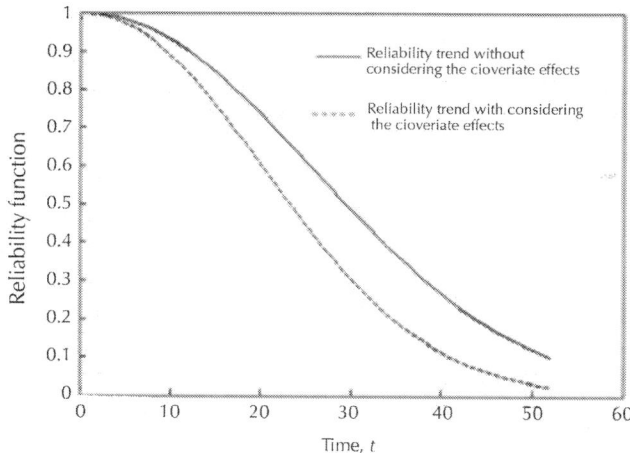

Figure 6: Effect of covariates on the Weibull reliability function.

The results presented in Table 6 and Fig. 4, Fig. 5 and Fig. 6 shows the respective comparisons of FTM for the case of the transmission belt with and without considering the covariate effects. By performing the FTM and considering the effects of external factors (covariate effects), some failure characteristics of the targeted components, such as MTTF, hazard rate, probability density and reliability functions, are more reliable. Therefore, this method is useful for better PM strategy development. Many researchers, such as Labib [35] and Tam et al. [36], regarded the need for considering the effects of external factors in FTM for effective maintenance strategy. Overall, the proposed failure analysis methodology, which integrates the FMECA and FTM based on PHM, provides a systematic process of how a better and more reliable failure analysis can be carried out. This methodology not only shows the identification and codification of external factors (covariate effect), but also presents the classification process of failure time towards censored and uncensored data based on criticality index estimation.

CONCLUSIONS

The integration of the FMECA and FTM based on PHM offers a practical and effective process to obtain more reliable results in failure analysis. FMECA is applied to identify and classify the physical failure characteristics of the targeted components, such as failure mode, cause, effect, and criticality index. The output of FMECA is then used in FTM. FTM based on PHM is used to determine more reliable failure characteristic parameters, such as MTTF, hazard rate, probability density, and reliability functions of the targeted component, because it considers external factors. Based on these results, the planned preventive strategies to address the targeted components can be more effective. From the research presented in this paper, a further study is being planned, which aims to validate the proposed methodology using more complex cases, such as the component/system with more than two failure modes, repairable type component, and multi-component structures. To deal with these cases, the authors believe that some modification on the proposed methodology may be required. However, the integration of FMEA and FTM based on PHM as the main structure of the methodology can still be retained.

REFERENCES

1. Nowlan FS, Heap HF. Reliability-centered maintenance. Springfield, VA: US Department of Commerce; 1978.
2. Badía FG, Berrade MD. Optimum maintenance of a system under two types of failure. Int J Mater Struct Reliab 2006;4(1):27–37.
3. SS-EN 13306. Swedish Standards Institute. Maintenance Terminology; 2001.
4. Moustafa MSA, Maksoud EY, Sadek S. Optimal major and minimal maintenance policies for deteriorating systems. Reliab Eng Syst Safe 2004;83:363–8.
5. Blischke WR, Murthy DNP. Case studies in reliability and maintenance. USA: John Wiley & Sons; 2003.
6. Wang H, Pham H. Some maintenance models and availability with imperfect maintenance in production systems. An Oper Res 1999;1999(91):305–18.
7. LaCombe D. Reliability control for electronic systems. New York, NY: Marcel Dekker; 1999.
8. Varzakas T, Arvanitoyannis IS. Application of failure mode & effect analysis (FMEA), cause & effect analysis and Pareto diagram in conjunction with HACCP to a strudel manufacturing plant. Int J Food Sci Technol 2007;2007(42):1156–76.
9. Ebeling CE. Reliability and maintainability engineering. United States of America: McGraw-Hill Companies, Inc.; 1997.
10. MIL-HDBK-338-IA. Military handbook – electronic reliability design handbook. Department of Defense, Washington, DC; 1988.
11. MIL-STD-1629A. Military standard – procedures for performing a failure mode, effects and criticality analysis. Department of Defense, Washington, DC; 1980.
12. Becker JC, Flick G. A practical approach to failure mode, effects and criticality analysis (FMECA) for computing systems.

In: High-assurance syst eng workshop, IEEE, Niagara on the Lake, Ont., Canada; 1996. p. 228–36.
13. Teng SH, Ho SY. Failure mode and effects analysis: an integrated approach for product design and process control. Int J Qual Reliab Manage 1996;13(5):8–26.
14. Xu K, Tang LC, Xie M, Ho SL, Zhu ML. Fuzzy assessment of FMEA for engine systems. Reliab Eng Syst Safe 2002;75:17–29.
15. Puente J, Pino R, Priore P, Fuente DDL. A decision support system for applying failure mode and effects analysis. Int J Qual Reliab Manage 2002;19(2):137–50.
16. Yang C-C, Lin W-T, Lin M-Y, Huang J-T. A study on applying FMEA to improving ERP introduction – an example of semiconductor related industries in Taiwan. Int J Qual Reliab Manage 2006;23(3):298–322.
17. Teng SG, Ho SM, Shumar D, Liu PC. Implementing FMEA in a collaborative supply chain environment. Int J Qual Reliab Manage 2006;23(2):179–96.
18. Arabian-Hoseynabadi H, Oraee H, Tavner PJ. Failure modes and effects analysis (FMEA) for wind turbines. Electr Power Energy Syst 2010;32:817–24.
19. Xiao N, Huang H-Z, Li Y, He L, Jin T. Multiple failure modes analysis and weighted risk priority number evaluation in FMEA. Eng Fail Anal 2011;18:1162–70.
20. Pickard K, Müller P, Bertsche B. Multiple failure mode and effects analysis: an approach to risk assessment of multiple failures with FMEA. In: Reliab maint symp annual. Piscataway: Institute of Electrical and Electronics Engineers Inc; 2005. p. 457–62.
21. Ferreira JLA, Balthazar JC, Araujo APN. An investigation of rail bearing reliability under real conditions of use. Eng Fail Anal 2003;10:745–58.
22. Liberopoulos G, Tsarouhas P. Reliability analysis of an automated pizza production line. J Food Eng 2005;69:79–96.

23. Tsarouhas PH, Varzakas TH, Arvanitoyannis IS. Reliability and maintainability analysis of strudel production line with experimental data – a case study. J Food Eng 2009;91:250–9.
24. Tsarouhas PH, Arvanitoyannis IS, Varzakas TH. Reliability and maintainability analysis of cheese (feta) production line in a Greek medium-size company: a case study. J Food Eng 94:233–40.
25. Tsarouhas PH, Arvanitoyannis IS. Quantitative analysis for peach production line management. J Food Eng 2011;105:28–35.
26. Oyebisi TO. On reliability and maintenance management of electronic equipment in the tropics. Tech 2000;20:517–22.
27. Martorell S, Sanchez A, Serradell V. Age-dependent reliability model considering effect of maintenance and working conditions. Reliab Eng Syst Safe 1999;64:19–31.
28. Dhillon BS, Liu Y. Human error in maintenance: a review. J Qual Maint Eng 2006;12(1):21–36.
29. Latorella KA, Prabhu PV. A review of human error in aviation maintenance and inspection. Int J Ind Ergo 2000;26:133–61.
30. Ahmad R, Kamaruddin S, Azid IA, Putra AI. Preventive replacement schedule: a case study at a processing industry. Int J Ind Syst Eng 2011;8(3):386–406.
31. Ghodrati B, Kumar U. Reliability and operating environment-based spare parts estimation approach – a case study in Kiruna Mine, Sweden. J Qual Main Eng 2005;11(2):169–84.
32. Coit DW, English JR. System reliability modelling considering the dependence of component environmental influences. In: Proc ann relia main symp, Washington, DC, USA; January 18–21, 1999. p. 214–8.
33. Cox DR. Regression models and life-tables. J Royal Stat Soc 1972;B34:187–220.
34. Kalbfleisch JD, Prentice RL. The statistical analysis of failure data. United State of America: John Wiley and Sons; 1980.
35. Labib AW. A decision analysis model for maintenance policy

selection using a CMMS. J Qual Main Eng 2004;10(3):191–202.
36. Tam ASB, Chan WM, Price JWH. Optimal maintenance interval for a multicomponent system. Prod Plan Cont Manage Oper 2006;17(8):769–79.

Chapter 2

The MRN Complex in Double-strand Break Repair and Telomere Maintenance

Brandon J. Lamarche[a], Nicole I. Orazio[a,b], and Matthew D. Weitzman[a]

[a]Laboratory of Genetics, The Salk Institute for Biological Studies, La Jolla, CA 92037, USA

[b]Graduate Program, Division of Biology, University of California, San Diego, CA 92093, USA

ABSTRACT

Genomes are subject to constant threat by damaging agents that generate DNA double-strand breaks (DSBs). The ends of linear chromosomes need to be protected from DNA damage recognition and end-joining, and this is achieved through protein–DNA complexes known as telomeres. The Mre11–Rad50–Nbs1 (MRN)

complex plays important roles in detection and signaling of DSBs, as well as the repair pathways of homologous recombination (HR) and non-homologous end-joining (NHEJ). In addition, MRN associates with telomeres and contributes to their maintenance. Here, we provide an overview of MRN functions at DSBs, and examine its roles in telomere maintenance and dysfunction.

INTRODUCTION

DNA double-strand breaks (DSBs) can be generated by chemical and physical damage inflicted by ionizing radiation, select chemotherapy drugs, and metabolic byproduct reactive oxygen species. DSBs can also result from errors during replication, and are produced by programmed enzymatic activities during meiosis and V(D)J recombination. Regardless of how they are generated, DSBs differ from all other types of DNA lesions in that the sequence information requisite for guiding repair is no longer contained within a contiguous duplex molecule. If left unrepaired, DSBs are among the most deleterious DNA lesions, with the potential to generate chromosomal translocations, aneuploidy, and increased incidence of malignancy. The importance and centrality of DSB repair pathways during the course of evolution is demonstrated by conservation of the core components from yeast to mammals. There are two major competing pathways for DSB repair: homologous recombination (HR) and non-homologous end-joining (NHEJ). Although we do not fully understand the regulation of pathway choice, the relative extent to which these two pathways are employed depends on the cell type, the phase of the cell cycle in which the DNA damage is encountered, and also varies between species [1] and [2].

At least three distinct functionalities are required for repair of DSBs: detection of the damage, an ability to control the cell cycle and transcriptional programs in response to the damage, and mechanisms for catalyzing repair of the lesion. The Mre11–Rad50–Nbs1 (MRN) complex sits at the hub of the eukaryotic DSB response mechanism, and has emerged as a crucial player in each of these three facets of DSB repair. This complex of proteins acts as DSB sensor, co-activator of DSB-induced cell cycle checkpoint signaling, and as a DSB repair effector in *both* the HR and NHEJ pathways [3], [4], [5], [6] and [7]. The MRN complex has also been found to associate with telomeres at the ends of linear chromosomes, where it contributes to their maintenance. Since MRN promotes sensing and repair of DNA ends, its presence at chromosome termini appears paradoxical. In this review, we first provide an overview of MRN's constituents, structure, catalytic activities, protein-binding partners, and signal transduction roles in the context of DSB repair within non-telomeric regions of the chromosome. We then discuss the emerging roles of the MRN complex in telomere maintenance and dysfunction.

THE MRN COMPLEX

Mre11, Rad50, and Xrs2 (the *Saccharomyces cerevisiae* homolog of vertebrate-specific Nbs1) were first identified via screens for yeast genes involved in meiotic recombination, and resistance to DNA damage induced by UV light and X-rays. Consistent with the nearly identical phenotypes resulting from defects in these three genes, Ogawa and co-workers demonstrated that Mre11, Rad50, and Xrs2 belong to the same epistasis group [8], and subsequently these proteins were shown to associate physically with each other in both yeast and mammals [9], [10] and [11]. Here, we describe specific features of each of the main components of the complex.

Mre11

Mre11 is a highly conserved 70–90 kDa protein composed of an N-terminal Mn^{2+}/Mg^{2+}-dependent [12]phosphoesterase domain, and two distinct C-terminal DNA-binding domains [13] and [14] (Fig. 1A). Isolated Mre11 forms stable dimers [15] that possess a number biochemical activities including: (i) intrinsic DNA binding activity [9], [16] and [17] with the specific ability to synapse DSB termini [15], and (ii) endo- and exonuclease activities against a variety of single-stranded DNA (ssDNA) and double-stranded DNA (dsDNA) substrates [17] and [18]. While these nuclease activities contribute to both NHEJ [19] and HR[20], it should be noted that Mre11 conspicuously lacks the 5'→3' exonuclease activity requisite for generating the long 3' ssDNA overhangs necessary for HR. Although it is possible that a protein-binding partner might switch the polarity of the exonuclease activity of Mre11, it is more likely that Mre11 facilitates the activity of additional DSB processing factors. In both yeast and mammals a number of 5'→3' exonucleases have been identified [12] and [21] that are capable of contributing to the generation of 3' overhangs during HR, giving weight to the idea that other enzymes act in concert with MRN [22].

In vivo, Mre11 exists in $Mre11_2Rad50_2$ "core" complexes where each Mre11 molecule binds a single Rad50 (Fig. 1B). The Nbs1 or Xrs2 proteins bind this core complex via interactions with Mre11 to give an overall stoichiometry of $Mre11_2Rad50_2Nbs1_2$, although there is some discrepancy over the number of Nbs1 proteins bound to the MR complex [11] and [23]. Supporting the importance of these interactions to MRN complex stability and function, Mre11 mutations that destabilize the MRN complex result in significantly decreased levels of Rad50 and Nbs1 in vivo [17] and [24], and knockdown of individual components of MRN can produce decreases in the other two members [25]. In reconstitution studies the addition of Rad50 enhances the affinity of Mre11 for DNA and stimulates its nuclease activity, and this is further enhanced by the addition of Nbs1 [17].

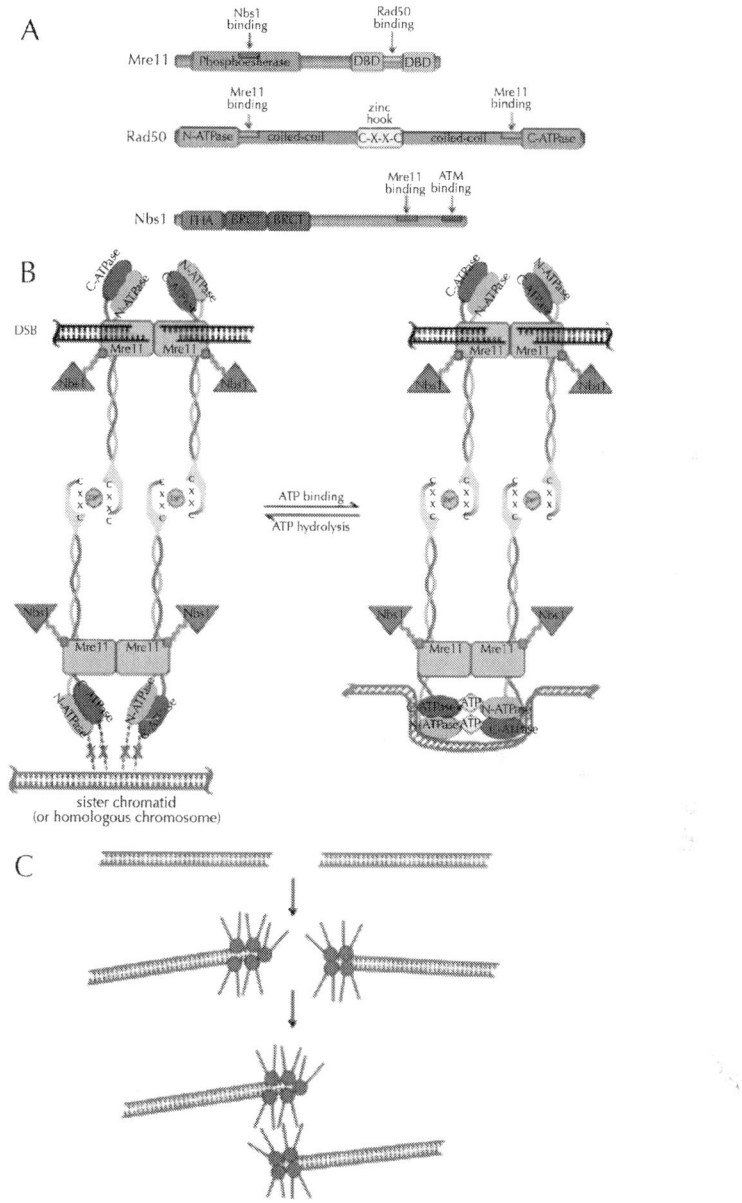

Figure 1: Characteristics of the MRN complex. (A) Domain organization of Mre11, Rad50, and Nbs1. (B) Model of intermolecular interactions within the MRN·DNA ternary complex. While there is evidence that multiple MRN complexes can cluster at DNA termini, for simplicity

only a single complex is depicted. See text for details. (C) Clustering of MRN complexes at DNA termini, and the subsequent tethering of multiple DNA molecules via the coiled-coil arms of Rad50. The DNA-binding globular heads of the MRN complexes are depicted by spheres.

Rad50

Rad50 is a ~150 kDa protein displaying both sequence and structural homology to structural maintenance of chromosome (SMC) family members that control the higher-order structure and dynamics of chromatin. The N-terminal Walker A and C-terminal Walker B nucleotide binding motifs stably associate with one another to form a bipartite ATP-binding cassette (ABC)-type ATPase domain [26] and [27] that preferentially binds and partially unwinds dsDNA termini [28]. The intervening ~575 amino acids form an anti-parallel coiled-coil that spans ~500 angstroms and terminates with a zinc hook (CxxC) motif [29] (Fig. 1A and B). Formation of the stable $Mre11_2Rad50_2$ core complex is achieved by each unit of the Mre11 dimer binding a Rad50 molecule at the intersection of its globular and coiled-coil domains [27] and [28]. This results in a spatial juxtaposition of the DNA-binding/termini-unwinding capacities of Rad50 with the DNA-binding, DSB tethering, and nuclease activities of Mre11 (Fig. 1B).

Independent of the Mre11-mediated dimerization of Rad50, biochemical analyses indicate that isolated Rad50 is able to form robust dimers under certain conditions [30]. Consistent with this, a crystal structure demonstrated that the globular ATPase/DNA-binding domains from two Rad50 molecules can associate [26]. Importantly, this interaction is ATP-dependent, with two ATP molecules getting sandwiched within the Rad50–Rad50 interface (Fig. 1B). Hopfner and co-workers deemed ATP hydrolysis by this complex too inefficient to support motor or helicase functions, and therefore proposed a "switch" function instead [26]. This "switch hypothesis" predicts that within the context of the $Mre11_2Rad50_2Nbs1_2$ complex, the two Rad50 molecules toggle between states in which their ATPase/DNA-binding domains are associated or disassociated (Fig. 1B). Since ATP binding-induced structural changes dramatically enhance the affinity of Rad50 for linear double-stranded DNA [26], an ATP binding/hydrolysis cycle may be a mechanism for modulating the length of time the MRN complex remains bound to a substrate or product.

Scanning force microscopy has demonstrated that while the globular head of the Mre11$_2$Rad50$_2$ complex associates with the termini of linear dsDNA, the two coiled-coil regions of Rad50 are flexible "arms", and project outward away from the DNA [28] (Fig. 1B). With increasing concentration, Mre11$_2$Rad50$_2$ complexes oligomerize at dsDNA termini, where the coiled-coil arms of Mre11$_2$Rad50$_2$ complexes mediate interactions between DNA termini [28] (Fig. 1C). Since this tethering of DNA termini was not observed at lower protein concentrations where only a single Mre11$_2$Rad50$_2$ complex was associated with a given DNA terminus, it was suggested that this may be a mechanism for achieving a robust and specific DSB tethering effect through the use of multiple weak interactions [28]. Crystallographic studies subsequently demonstrated that it is the zinc hook of Rad50 that mediates interaction between the coiled-coil arms of Rad50 molecules [29] (Fig. 1B). Mutation of the zinc-coordinating cysteine residues of Rad50 resulted in significant DSB repair defects in yeast, and complete replacement of the zinc hook with an FKBP dimerization cassette mitigated this repair defect [31]. This suggests that the ability of Rad50 molecules to bind one another via the distal terminus of their coiled-coil arms is important for MRN function.

The available structural and functional data indicate that a major role of the MRN complex is to mediate spatial juxtaposition of DNA molecules, and that this involves two distinct modes of binding. In the current model, the Mre11 dimer facilitates short-range synapsis of the two ends of a DSB, while Rad50 enables long-range tethering of two DNA molecules (such as a broken chromosome and its sister chromatid) by dimerizing via its zinc hook (Fig. 1B) [15] and [29]. In its fully extended conformation, the zinc-mediated Rad50 dimer places the two DNA molecules being bound by each Rad50/Mre11 globular head ~1200 angstroms from each other [29]. Importantly, however, the coiled-coils of Rad50 are extremely flexible, suggesting that even while they remain tethered by Rad50 the sister chromatids would be capable of achieving a tighter spatial proximity more conducive to homology-mediated repair.

Nbs1

The third member of the MRN complex is Nbs1, a 65–85 kDa protein. Nbs1 consists of an FHA domain and two adjacent BRCT domains at its N-terminus, in addition to an Mre11-interaction domain at its C-terminus[32] and [33] (Fig. 1A and B). The FHA domain binds phosphorylated threonine residues in Ser-X-Thr motifs present in DNA damage proteins, including Mdc1 and Ctp1. The BRCT domains in human Nbs1 bind Ser-X-Thr motifs when the serine residue is phosphorylated. These phospho-dependent interactions are important for recruiting repair and checkpoint proteins to DNA breaks [32] and [33]. By virtue of its nuclear localization signal and its interaction with Mre11, Nbs1 is responsible for translocation of the MRN complex into the nucleus. This can be observed with Nbs1 mutants lacking the Mre11-binding domain, which are themselves nuclear, while the $Mre11_2Rad50_2$ core complex remains cytoplasmic [34].

Limited proteolysis of Nbs1 yields a stable N-terminal fragment and C-terminal degradation products, suggesting that the FHA/BRCT core of Nbs1 is linked to Mre11 via a flexible tether [33] (Fig. 1B). While Nbs1 stimulates the DNA binding and nuclease activities of the MR complex [17], it does not itself possess a known enzymatic activity. Rather, Nbs1 contributes to DSB repair primarily by mediating protein–protein interactions at DNA breakage sites. The central region of Nbs1 possesses several SQ motifs that are phosphorylated by the ATM kinase as part of the DNA damage response. The C-terminus also contains a domain that interacts with ATM and recruits it to DSBs [35] and [36], and is required for the induction of apoptosis in response to damage [37].

MRN MUTATIONS: HUMAN DISEASES AND MOUSE MODELS

In humans, hypomorphic mutations in the *NBS1* gene result in Nijmegen breakage syndrome (NBS)[38] and [39], a rare autosomal recessive disorder characterized by microcephaly, immunodeficiency, and predisposition to cancer [40]. The most common mutation in NBS patients is 657del5, which results in truncated proteins that partially maintain some functions of the full length protein: a short N-terminal fragment that includes the FHA/BRCT domains can be detected, and in some cell types there is also a 70 kD C-terminal fragment which is capable of interacting with Mre11. Hypomorphic mutations in human*MRE11* lead to ataxia-telangiectasia-like disorder (A-TLD) [24], in which patients display ataxia and neurodegeneration, resembling the phenotypes of ATM deficiency [41]. A single patient has been reported with a hypomorphic mutation in *RAD50*, who exhibited phenotypes similar to NBS [42]. Patients with NBS are prone to developing malignancies such as lymphoma and leukemia [40], and somatic alterations in MRN may contribute more widely to carcinogeneisis [43]. Cell lines derived from NBS and A-TLD patients have been valuable in dissecting the functions of MRN and the consequences of compromised function. Consistent with the involvement of MRN in cell cycle checkpoint signaling and DNA repair, cells from patients with NBS and A-TLD display increased radio-sensitivity and are defective for checkpoint activation[24] and [38].

Animal models of MRN defects have been useful in understanding the pathology of human disease. Null mutations in members of the MRN complex lead to embryonic lethality in mice, and therefore mouse models have been generated to mimic the hypomorphic mutations identified in human. Models have been made for A-TLD (Mre11$^{ATLD1/ATLD1}$) [44] and NBS (by generating truncated proteins NBS$^{b/b}$ and NBS$^{m/m}$, as well as a humanized NBS syndrome mouse model hNBS$^{657\ 5}$) [45] and [46]. In addition, insights have been gleaned from mice engineered to express either Nbs1 mutants lacking specific domains[37] and [47] or the Rad50S hypermorphic mutation initially described in *S. cerevisiae* (Rad50$^{S/S}$) [48]. While the disease phenotypes of these mouse models are less severe than those in the corresponding human disorders, together with cell and tissue specific knockouts, they can individually recapitulate various aspects of the human disease including immunodeficiency, cancer predisposition, and germline and neuronal defects [46], [49] and [50].

Immunological deficiency, including defects in both humoral and cellular immunity, is a hallmark of human NBS. Specifically, human B-cells display decreased variability in IgG and IgA subtypes (compared to the IgM subtype), and an increased susceptibility to lymphogenesis. Conditional mouse models with targeted deletion of *NBS* in B lymphocytes revealed that the decreased immunoglobulin variability is due to a defect in class switch recombination (CSR) [51]. Since CSR requires repair of the DSBs induced as intermediates in the process, the basis for CSR defects in MRN deficient cells is likely due to the functions of MRN in NHEJ, although signaling defects could also contribute.

Cancer predisposition and chromosomal instability in NBS patients appears in the form of lymphomas, particularly B-cell lymphomas. NBS, A-TLD, and Rad50$^{S/S}$ mice display increased tumorigenesis in a p53 null background, while *Nbs1* heterozygotes are susceptible to various types of cancers independent of p53 [46] and [52]. Cells derived from the model animals are sensitive to ionizing radiation, defective for checkpoint activation and display increased amounts of chromosomal aberrations [45], [48], [52] and [53]. Germ line defects in NBS patients include infertility and compromised sexual maturation. The hNBS$^{657\ 5}$ mouse recapitulates these effects when compared to littermate controls [46]. The male mice have smaller testes with histological degeneration, increased apoptosis, and delayed appearance of germ cells. The adult female mice fail to breed and have small ovaries devoid of oocytes. Analysis of meiotic chromosome spreads in oocytes at birth reveals depletion in the diplotene stage, suggesting that Nbs1 is required for meiotic progression. Analysis of meiotic events in mice harboring the hypomorphic Mre11 and Nbs1 mutations also revealed defects in synapsis of homologous chromosomes and crossovers, and suggested that MRN contributes to normal sex-specific differences in meiosis [54].

Neurological defects are a hallmark of MRN mutation, with microcephaly and ataxia observed in NBS patients, and neurodegeneration detected in A-TLD patients [55]. These different neuropathies are probably reflective of the fact that the respective disease-causing mutations differentially impact DNA damage signaling in the brain. Conditional disruption of the murine ortholog of the human *NBS1* gene in the mouse central nervous system (CNS) causes ataxia, microcephally, cerebellar disorganization and disruption of the visual system [50] and [56]. Although the hypomorphic NBS and A-TLD mice did not exhibit neurological defects, when damage was introduced during development these mice displayed distinct neurological phenotypes which were attributed to differential activation of an apoptotic response [44], [52] and [57]. The MRN complex is required for activation of the ATM-dependent p53 apoptotic response either during neural development, or under conditions of DNA damage [56] and [57]. NBS mice expressing the C-terminal domain of Nbs1 retain the ability of Nbs1 to interact with Mre11 and ATM, and this is sufficient to activate an ATM-dependent DNA damage response [35], [36] and [37], which can lead to apoptosis during neuronal development and result in microcephaly [57]. In contrast, A-TLD mice subjected to ionizing radiation do not exhibit normal ATM signaling, p53 activation, and apoptosis. Therefore, accumulation of mutations and genomic damage in these cells may be responsible for the neurodegeneration seen in A-TLD patients [57].

THE MULTIPLE ROLES OF MRN IN DSB REPAIR

The MRN complex plays critical roles in pathways involved in DNA damage repair, checkpoint activation, telomere maintenance, meiosis and DNA replication. As a prelude to our discussion of MRN functions in telomere maintenance, here we focus on the activities of the complex most relevant to detection and repair of DSBs. We refer the reader to excellent recent articles that review additional aspects of MRN function in meiosis, replication and checkpoint signaling [13], [58], [59] and [60].

MRN-mediated DSB Detection and Activation of Signal Transduction

Coordination of the DNA damage surveillance and repair systems functions to prevent transmission of genetic mutations [61]. Sensors detect damaged DNA and activate protein kinases that launch a network of signal transduction cascades that form the DNA damage response (DDR) [62]. This network prevents an array of human diseases, and when compromised it can lead to genomic instability and cancer [61]. The primary signaling kinases are the ataxia telangiectasia mutated protein (ATM), the ATM and Rad3-related kinase (ATR), and the DNA-dependent protein kinase (DNA-PK). When these DNA damage kinases are activated they phosphorylate a specific serine residue of histone H2AX at the breakage site and flanking chromatin [63] and [64]. Proteins involved in repair and checkpoint activation subsequently accumulate at the DSB to form foci that are visible by fluorescent microscopy. Post-translational modifications to histones and repair proteins determines the temporal order of accumulation at the sites of damage [65] and [66]. Recruited proteins are involved in signal amplification, chromatin modification, and repair of the DSB. The kinase pathways induced by DSBs in human cells result in phosphorylation of more than 700 different proteins, including the mediators Mdc1 and 53BP1, downstream checkpoint kinases Chk1 and Chk2, and proteins with a diverse array of functions [67]. Signaling in response to DNA damage activates three distinct arrests to cell cycle progression: the G1/S checkpoint prevents cells from entering S phase, the intra-S checkpoint inhibits replication during S phase, and the G2/M checkpoint prevents damaged cells from entering mitosis.

Checkpoint activation can be accompanied by changes in cellular transcription profiles and apoptotic death pathways.

The MRN complex plays a role very early in the DDR, acting as a sensor of DSBs [68]. It has been demonstrated that mutation, knockdown, degradation, or mislocalization of MRN components leads to defective ATM signaling [24], [69], [70] and [71]. In addition to being present diffusely throughout the nucleus [72], a fraction of the MRN complex is also sequestered within sub-nuclear compartments known as promyelocytic leukemia (PML) bodies [73]. Consistent with the role of MRN as a DSB sensor, the concentration of MRN components remains constant throughout the cell cycle [72]. Immunofluorescence (IF) analyses indicate that upon DSB formation, the MRN complex rapidly relocalizes from both the diffuse pool and PML bodies to breakage sites [73] and [74]. MRN can be detected within these repair foci using standard IF because hundreds, if not thousands, of copies of the protein complex accumulate within the vicinity of the DSB [75]. Rapid recruitment of fluorescently labeled Nbs1 has also been demonstrated in live human cells with DSBs generated by laser micro-irradiation, which is restricted to small sub-nuclear areas[76]. Potential benefits of such rapid detection are that it minimizes (i) the time that free DNA termini remain vulnerable to non-specific degradation, and (ii) the amount of time that DSB termini are able to diffuse away from one another. In the presence of multiple DSBs, the latter would serve to maximize the probability of MRN synapsing the two termini that were originally contiguous.

In undamaged cells, ATM exists as a homodimer that is incapable of phosphorylating its substrates [77], presumably because dimerization occludes substrate binding. Upon induction of DSBs by ionizing radiation ATM is recruited to DSBs, at least in part, via a direct interaction with a C-terminal motif in Nbs1[35] and [36] (Fig 1A). This evolutionarily conserved C-terminal region of Nbs1 may not be absolutely required for recruitment and activation of ATM [37] and [47], since its loss appears to be compensated for by mediators such as 53BP1 [78]. By monitoring laser micro-irradiation-induced DSBs in real time, it was demonstrated that an MRN-dependent accumulation of ATM occurs within seconds [79]. This MRN–ATM interaction, which is optimal in the presence of dsDNA termini [35] and [80] and is stimulated by Rad50-mediated melting of the duplex terminus [80], increases the effective concentration of ATM in the vicinity of the DSB and also promotes the autophosphorylation of ATM dimers [80]. Autophosphorylation of ATM at serine 1981 triggers dissociation of the inactive dimer into kinase active monomers [77]. Autophosphorylation of ATM is essential to the DSB response not only because of the subsequent dissociation to active monomers, but also because it enables ATM to be retained at the breakage site [81]. The S1981A mutant of human ATM, which forms dimers that cannot be activated via autophosphorylation, is initially recruited to DSBs at a rate similar to that of wild-type ATM but is released from the break [79], and is consequently incapable of catalyzing essential downstream phosphorylation events [82].

At low doses of radiation the presence of functional MRN complex enhances activation of ATM, although there are also MRN-independent mechanisms of ATM activation. Treatment of cells with mild hypotonic solution or with the topoisomerase II inhibitor chloroquine, both of which induce chromatin structural changes without inducing DSBs, triggers ATM activation to a degree similar to that achieved by ionizing radiation [77]. Moreover, it was recently shown that modification of chromatin structure by inhibition of histone deacetylases can also activate ATM [83] and [84]. Activation of ATM via these types of chromatin modification (in the absence of DSBs) results in ATM-mediated phosphorylation of p53 [77] but not of proteins such as Nbs1 or SMC1 that are integral to the DSB coping mechanism [82]. This raises the possibility that upon DSB formation, ATM is actually responding to a change in the higher order structure of chromatin. In addition to MRN, other proteins recruited through ATM interactions, such as Tip60 [85], may contribute to ATM activation and modification of chromatin at DSBs. Stimuli that alter chromatin and induce ATM activation in the absence of DSBs may also involve additional ATM-binding proteins, such as ATMIN, that could compete with Nbs1 [86]. In addition to recruitment of ATM and activation of it's signaling, MRN also participates in the early steps of end resection at DSBs (see below) and this leads to subsequent activation of the ATR kinase [87]. Processing of DSBs is required to generate the substrates that lead to ATM-dependent activation of the ATR kinase [88]. It is proposed that ATM activation leads to resection of the DNA ends at breaks [87], and that production of single-stranded tails transforms the ends from ATM substrates into ATR substrates [88]. In support of this model, the nuclease activity of Mre11 has been shown to contribute to ATR activation in a mouse model [89]. In response to other types of DNA damage, functional MRN complex also promotes ATM-independent activation of the ATR kinase, although the mechanism by which this occurs is not clear[25], [90], [91], [92] and [93]. It has also been shown in *Xenopus Laevis* extracts that MRN-dependent processing of DSBs leads to the accumulation of short single-stranded DNA oligonucleotides that stimulate ATM activity [94], and these could have relevance for ATR signaling.

Activation of ATM and ATR at DSBs initiates a signaling network that (i) provides regulation of the cell cycle (via the protein kinases Chk1 and Chk2) [95], (ii) promotes chromatin remodeling that is necessary for allowing the repair machinery access to the lesion [81] and [96], and (iii) contributes to the recruitment and retention of additional proteins responsible for repairing the break. The DSB-induced signaling cascade that mediates checkpoint activation and lesion repair involves a large number of proteins and post-translational modifications (phosphorylation, acetylation, methylation, ubiquitinylation, sumoylation) that have been recently reported and reviewed elsewhere [66], [85], [95], [97], [98], [99] and [100].

Repair Pathway Selection: HR versus NHEJ

The relative extent to which DSBs are repaired via HR versus NHEJ varies among different species and cell types. In yeast and the simpler eukaryotes, which possess compact genomes and a relative paucity of repetitive sequences, HR makes a greater contribution to DSB repair. However, in mammals, where intergenic regions are larger and repetitive regions more abundant, it has been suggested that NHEJ is faster and more efficient [101]. This predisposition towards NHEJ in mammals may reflect the fact that gross chromosomal rearrangements can arise if the wrong region is utilized during HR, in a templating molecule containing multiple regions of repetitive sequence [102]. Although a given organism may display a general preference for HR or NHEJ, the extent to which each pathway is employed temporally fluctuates depending on the phase of the cell cycle [103].

There are multiple HR sub-pathways [104], but the defining feature of HR in mitotic cells is that it utilizes the sister chromatid to guide repair of a DSB during the S and G2 phases of the cell cycle [2]. In meiotic cells, HR utilizes either the homologous chromosome or the sister chromatid to guide repair during the first and second meiotic divisions, respectively [105]. During HR, exonucleolytic processing generates long 3' ssDNA overhangs on each side of the DSB which invade and base pair with the homologous regions of the intact sister chromatid (or homologous chromosome), and prime DNA synthesis [22]. Resolution of this crossed over complex yields two intact chromosomes of identical sequence, ensuring high fidelity during the HR repair process.

In contrast to HR, NHEJ is suppressed during meiosis and is mainly employed during the G0, G1, and early S-phases of the mitotic cell cycle, when sister chromatids are not present to guide repair [1]. During NHEJ the ends of the break are joined irrespective of their sequence, and this pathway is therefore inherently error-prone. In NHEJ repair the two termini of the DSB are either directly ligated in classical NHEJ (C-NHEJ), or distal regions of microhomology (consisting of 1–4 nucleotides) on each side of the DSB are utilized to align the fragments prior to ligation during the alternative NHEJ (A-NHEJ) pathway (Fig. 2A and B). In both types of NHEJ, processing of the termini via nuclease-mediated removal of nucleotides or polymerase-mediated gap filling may be employed. This ability to repair DSBs whose termini display little or no homology is both the strength and the weakness of NHEJ, in that a lethal DSB is traded for a small deletion or insertion. In the presence of multiple DSBs, NHEJ repair can give rise to gross chromosome rearrangements, such as inversions and translocations [106], because there is no mechanism for determining which of the multiple DNA termini were originally contiguous.

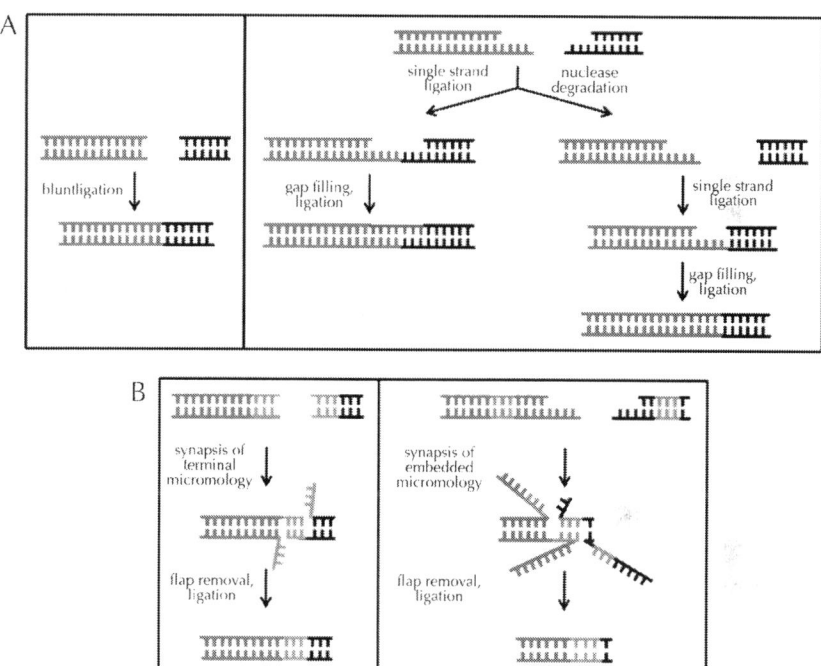

Figure 2: Models of NHEJ. (A) C-NHEJ, also known as direct end joining, is carried out with minimal processing of the DSB termini prior to ligation. Both blunt and protruding termini are amenable to C-NHEJ. A single DSB can be processed via multiple routes, with the product sequence being determined by the chronology in which ligase, nuclease, and polymerase activities are employed. Blue regions denote gaps filled in by a DNA polymerase. (B) A-NHEJ involves modest resection of DSB ends (<100 nucleotides) until regions of microhomology are encountered which can guide reattachment of the DNA ends. Regions of microhomology are depicted in green.

Resection at DSBs plays a central role in determining the outcome of the competition between HR and NHEJ. During the S and G2 phases of the cell cycle, DSBs are resected to give extensive 3' ssDNA overhangs on each side of the break [87], [107] and [108]. This serves to generate a substrate for the HR-specific ssDNA-binding factors RPA and Rad51, and thus ensures that the DSB will be repaired via HR[106]. It also reduces the efficiency of NHEJ because the NHEJ-specific DNA termini binding factor Ku70/80 has poor affinity for ssDNA [109]. In contrast, during the G0 and G1 phases, when sister chromatids are not present to facilitate HR, the cellular DSB resection activity is downregulated [110], which gives the NHEJ machinery opportunity to bind and process the break.

The Contribution of MRN to Repair Pathway Selection

The MRN complex, via its DNA end-processing activities, plays a pivotal function in initiating the processes that direct a DSB down the most appropriate repair pathway [22] and [106]. Although the MRN complex plays a key role, it is not sufficient and requires collaboration with other factors. The exonuclease activity of Mre11 in vitro operates in the opposite polarity to that required for HR resection in vivo. This has prompted several groups to search for additional factors that work with MRN to facilitate DSB processing. In budding yeast *S. cerevisiae* the Sae2 protein is required in conjunction with MRX for processing meiotic DSBs and promoting resection [111]. Sae2 has been suggested to be an endonuclease that cooperates with Mre11 to cleave certain DNA structures [112] and process covalent protein-linked DSBs [113]. Ctp1 in the fission yeast *Schizosaccharomyces pombe* [114] and CtIP in human cells [115] and [116] have been proposed to be functional counterparts of Sae2 for resection and repair pathway choice. Human CtIP undergoes DSB-induced phosphorylation, localizes to DSB repair foci, directly binds the MRN complex, and catalyzes or confers upon MRN the ability to catalyze 5'→3' resection at DSBs [117]. By binding to DSBs after ATM activation, CtIP appears to facilitate the transition from DSB sensing to end-processing [118]. CtIP depletion results in attenuated recruitment of RPA and ATR to damage induced by laser micro-irradiation[116]. Since RPA binds the extensive 3' single-stranded overhangs generated by DSB resection, and ATR is the protein kinase that signals the presence of ssDNA, these data were used to suggest that CtIP is integral to the DSB resection process [116]. This is consistent with the reduction in HR frequency observed with CtIP depletion. Simultaneous depletion of both CtIP and Mre11 reduces HR frequency to the same degree as CtIP depletion alone [116], suggesting that these proteins function within the same pathway. In further support of this, deletion of the C-terminal region of CtIP that mediates its interaction with MRN abrogates the ability of CtIP to promote ssDNA formation [116]. Important questions remain relating to the specific role played by MRN/CtIP during the DSB resection step of HR in mammalian cells. Human CtIP shares only a very small stretch of homology with Sae2 and it has not yet been demonstrated whether CtIP has nuclease activity.

After initial processing by a complex containing MRN and CtIP, further resection by the combined action of other helicases and nucleases generates the large regions of ssDNA required to complete the HR pathway[22] and [106]. In *S. cerevisiae*, MRX/Sae2 catalyzes the initial removal of a few hundred nucleotides from a DSB [119] and [120] and subsequently the processive 5' 3' ExoI exonuclease or the Dna2 exonuclease (in conjunction with the Sgs1 helicase) continue resection to give kilobase-sized 3' ssDNA overhangs[119] and [120]. DSB resection in vertebrates likely proceeds via a similar mechanism where resection is initiated by MRN/CtIP and then completed by ExoI in conjunction with the BLM helicase [121]. Upon DSB induction, specific damage signals propagate outwards from the break along chromatin, raising the possibility that it is these modifications (and the resultant alteration of chromatin structure) that restrict the MRX/Sae2- and ExoI-mediated phases of resection to ~100 nucleotides and ~2–3 kb, respectively.

The mechanisms responsible for the cell cycle dependence of the DSB resection activity are starting to come into focus. The ability to direct DSBs towards HR during S/G2 but not during G1 appears to be rooted in both the cellular concentration of CtIP and its phosphorylation state [117]. Consistent with CtIP's role in the DSB resection step of HR, and the minimally detectable DSB resection activity seen in G1 extracts, CtIP is barely detectable during G0/G1 but displays maximal concentration during S/G2 [122]. A similar situation is seen for the *S. pombe* Ctp1 protein [114]. In chicken cells, phosphorylation of CtIP specifically in S/G2 is essential for its function in the DSB resection step of HR [123]. It has recently been shown in *S. pombe* that recruitment of Ctp1 by the N-terminus of Nbs1 involves phosphorylation [33] and CK2 has been suggested to mediate this phospho-dependent interaction [124]. Phosphorylation may also play a role in recruitment of CtIP in humans [115] and Sae2 in budding yeast [125], and in both cases phosphorylation mediated by CDK provides an explanation for the cell cycle control of DSB resection.

The extensive repertoire of structural and catalytic functions attributed to MRN is utilized differently within the HR and NHEJ pathways. The initial contribution of MRN to HR is that of DSB sensing. In contrast, a unique set of NHEJ proteins are employed for detecting, synapsing, and processing DSBs. Similar to MRN, these NHEJ repair factors are present and active throughout the cell cycle [126]. It is presently unclear how MRN out-competes the NHEJ machinery for its place at DSB termini during the S and G2 phases of the cell cycle. A significant body of data indicates that Ku suppresses HR [127], [128] and [129] and reduces DSB resection during HR [130], while loss or mutation of Ku increases the frequency at which DSBs are processed via the HR pathway [131]. These data support a model in which Ku and MRN, simultaneously present throughout the cell cycle, compete for DSB termini. This raises the issue of whether the initial DSB binding event is completely random or alternatively whether there are mechanisms that promote Ku binding during G1 and MRN binding during S/G2. If MRN binds a DSB during G1, 5'→3' resection is prevented by both the phosphorylation state and low concentration of MRN-associated CtIP. Whether an analogous mechanism for suppressing NHEJ if Ku binds a DSB during S/G2 remains to be determined. If Ku or MRN binds a DSB during an "inappropriate" phase of the cell cycle, having a mechanism in place for actively removing them would appear to be important. Indeed, it was recently shown that removal of Ku from an unrepaired DSB is dependent on functional MRN, and requires the ATP-binding function of Rad50[132].

Repair Functions of MRN in HR and NHEJ

The crystal structure of the Mre11 dimer bound to two DNA termini suggests that a single MRN complex spatially juxtaposes the ends of a broken chromosome [15] (depicted in Fig. 1B). While this short-range tethering (i.e., synapsis) function could conceivably be useful in NHEJ, where termini undergo minimal processing prior to ligation, it is not obvious that this would be beneficial to HR where DNA ends are resected to give expansive tracks of ssDNA. It is unclear whether DSB termini must be synapsed in order to initiate HR 5'→3' resection or whether the termini can be bound and resected independently of one another.

Although Mre11 endows the MRN complex with both ssDNA endonuclease and dsDNA 3'→5' exonuclease capabilities, these nuclease activities do not have equal importance during HR. Williams and co-workers identified point mutations that abrogate either the exonuclease activity or both the exo- and endonuclease activities of Mre11 [15]. Studying these mutants in *S. pombe* indicates that while 3'→5' exonuclease activity is dispensable for HR, loss of endonuclease activity results in a severe HR defect approaching that observed for Mre11 knockout [15]. IR is capable of producing DSBs in which the 3' moiety does not contain the 3'-hydroxyl necessary for polymerase-mediated extension after strand invasion/base pairing. Therefore, the 3'→5' exo activity of Mre11 would appear to be ideal for removing these 3' "blocking groups". The dispensability of the 3'→5' exonuclease activity of Mre11 in this experiment suggests that either (i) the assay is not sensitive enough to detect defects in processing these 3' blocking groups, which may be present in only a fraction of IR induced DSBs, (ii) that the endonuclease activity of Mre11 can be utilized for cleaning up 3' termini, or (iii) that a different nuclease is employed in these situations. Evidence from a nuclease deficient Mre11 mouse model (Mre11$^{H129N/}$) suggests a role for the nuclease activity in early events at DSBs during HR in higher eukaryotes [89].

Initial analyses of the importance of MRN to NHEJ produced conflicting results [133] and [134], but emerging data have now firmly established roles for MRN in both C- and A-NHEJ. In C-NHEJ, broken termini are initially bound by the ring-shaped Ku70/Ku86 heterodimer. These Ku-bound termini recruit and activate the DNA-dependent protein kinase catalytic subunit (DNA-PK$_{cs}$), which phosphorylates repair factors and checkpoint proteins in a manner analogous to, but distinct from, that of ATM during HR[135] and [136]. While Ku and DNA-PK$_{cs}$ mediate synapsis of the two DNA termini, the Artemis nuclease cleans up the termini via both endo- and exonuclease activities [137] and [138]. Finally, the XRCC4-DNA ligase IV complex associates with the terminal complexes and catalyzes sealing of both strands. C-NHEJ, also known as direct end-joining, is carried out with minimal processing of the DSB termini prior to ligation (Fig. 2A). In the absence of C-NHEJ factors such as XRCC4, Ku, and DNA-PK$_{cs}$, the recently described alternative A-NHEJ pathway becomes more prevalent. A-NHEJ involves modest resection of DSB ends (<100 nucleotides) until regions of microhomology are encountered which can guide reattachment of the DNA ends (Fig. 2B). Multiple groups have demonstrated that deletion of the XRCC4 gene both decreases the efficiency of NHEJ and causes a shift from usage of C-NHEJ to utilization of A-NHEJ instead [5] and [7]. Sequencing the joints of repaired molecules can determine which pathway has been employed. Recent studies show that siRNA-mediated knockdown of Mre11 results in reduced end-joining efficiency in both XRCC4$^{+/+}$ and XRCC4$^{-/-}$ backgrounds [4] and [5]. This indicates that MRN functions in both the C-NHEJ and A-NHEJ pathways. The function of MRN in NHEJ appears to be independent of ATM signaling since these results can be reproduced in the presence of a chemical inhibitor of ATM [5]. Although C-NHEJ is used predominantly in an XRCC4$^{+/+}$ background, imprecise (i.e., deletion-prone) microhomology-mediated A-NHEJ products are also detected. Depletion of Mre11 in a mouse ES XRCC4$^{+/+}$ background reduces the frequency at which DSBs are repaired using microhomology, suggesting that Mre11 facilitates resection in the search for microhomologies distal to the break [5]. Nuclease-dead mutants can be used to determine whether the role of Mre11 in this C-NHEJ-competent background is due to tethering versus resection activities. In C-NHEJ-competent *S. cerevisiae*, end-joining defects associated with Mre11 deletion can be rescued by nuclease-defective Mre11, indicating a structural rather than catalytic function for yeast Mre11 in this context [139]. When the C-NHEJ system is compromised, as in the mouse ES XRCC4$^{-/-}$ background, knockdown of Mre11 results in a decrease in the length of resection tracks prior to end-joining [5]. Whether Mre11 is itself catalyzing resection in the search for microhomology or is simply facilitating this process has not been determined. Depletion of CtIP in asynchronous SV40 transformed human fibroblasts causes a significant decrease in end-joining efficiency [4], consistent with a facilitative role for MRN. These recent studies indicate that in contrast to the situation with HR, knowledge of the roles of MRN in NHEJ is just beginning to take shape. Among the important questions that will need to be addressed are how MRN facilitates repair, why it is not sterically occluded by the presence of the NHEJ machinery, and what the specific roles of Mre11 (and other nucleases) are in resection prior to end-joining.

MRN IN TELOMERE MAINTENANCE AND DYSFUNCTION

The evolutionary transition from circular to linear chromosomes brought with it two new challenges to genome stability. The first of these, known as the "end replication problem", relates to the loss of nucleotides from the 5' terminus of the lagging strand after every round of DNA replication. The second challenge to genome stability associated with linear chromosomes is that of preventing the chromosome termini from being recognized and processed as DSBs. The specialized repetitive sequences and protein–DNA complexes that comprise telomeres function both to maintain the chromosome ends and to provide protection from the DNA repair machinery.

Removal of the RNA primers from lagging strand Okazaki fragments results in a gap of missing nucleotides at the 5' terminus, which cannot be filled by DNA polymerases due to the strict 5' 3' polarity of their synthesis activity. Consequently, in the absence of a prophylactic mechanism, continuous cycles of replication would cause the genomes of individual organisms to grow progressively shorter and genes would be lost [140]. The addition of telomeres, non-coding repetitive DNA sequences, to the termini of chromosomes overcomes this problem. Telomerase, a specialized reverse transcriptase that carries its own RNA template, synthesizes telomeres of sufficient size to ensure that genetic information is not lost due to the end replication problem. After telomerase has synthesized a guanine-rich 3' single-stranded extension (composed of TTAGGG repeats in humans) at the chromosome terminus, the cytosine-rich complementary strand is synthesized by traditional semi-conservative replication. In yeast, telomerase is active at each round of DNA replication, thereby ensuring that telomere length maintains a steady state. In contrast, during vertebrate development and within stem cell populations, telomerase synthesizes telomeres with a length that is sufficient to sustain numerous future replication cycles. Since telomerase is inactive in somatic cells, telomeres grow progressively shorter over time until they reach a critical length, at which point cell senescence or apoptosis is triggered. Vertebrate telomeres therefore provide a solution to the end replication problem in a temporally finite manner. This brings the added benefit of tumor suppression, in that unchecked cellular replication associated with cancer is thwarted when these cells reach their critical telomere length.

Throughout Eukarya, telomeres are bound by specific proteins that sequester the free chromosome termini within a nucleoprotein cap [141]. In higher eukaryotes these same telomere-binding proteins additionally promote the formation of a unique lariat-like structure called the "t-loop", which provides a second layer of protection from the DSB repair machinery [140] and [142]. Critical to t-loop formation is the generation of a 3' single-stranded overhang (termed the "G-overhang" because it is present on the G-rich strand of the telomere) at each chromosome terminus. Since the newly synthesized lagging strand is missing a portion of its 5' terminus, a short 3' G-overhang (~10 nucleotides) is inherently present at this end of a newly synthesized chromosome. In contrast, leading strand synthesis generates a blunt chromosome end. In higher eukaryotes both this blunt end and the short G-overhang at the opposite chromosome end are processed by an unknown nuclease to generate mature G-overhangs 50–300 nucleotides in length [140]. Folding each chromosome terminus back upon itself enables the G-overhang to invade and base pair with the complementary strand (analogous to what occurs during HR), giving rise to the t-loop lariat (Fig. 3). Although G-overhangs are also present at telomeres in *S. cerevisiae*, they are only 12–14 nucleotides in length [143] and t-loops have not been observed. Regardless of whether or not t-loops are employed, the importance of sequestering telomeres into nucleoprotein caps is made clear by the fact that cap disruption can result in shortening or lengthening of telomeres, telomere fusion, telomere loss, elevated levels of recombination, and checkpoint activation [144].

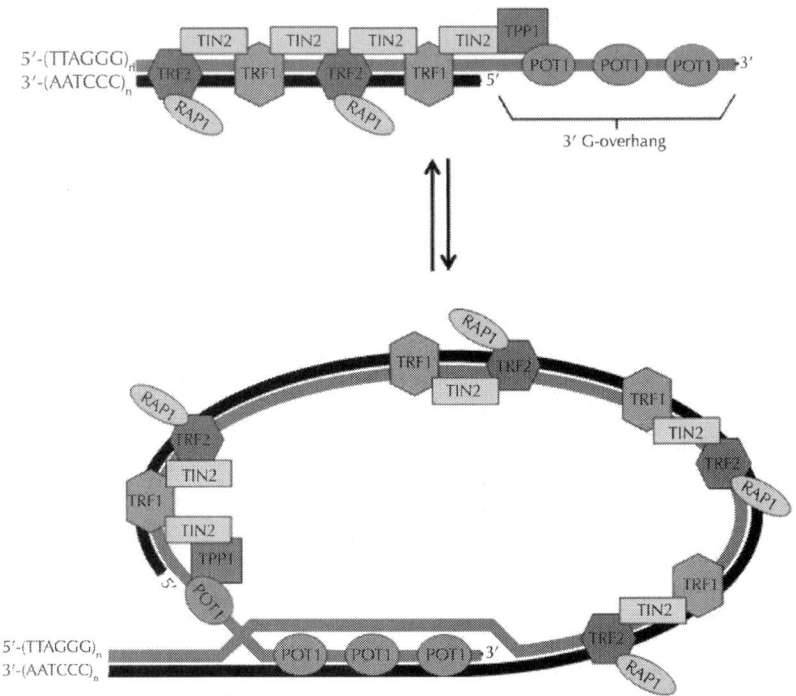

Figure 3: Human telomere structure. Specific interactions between shelterin components, and between these proteins and specific regions of telomeric DNA are highlighted. Looping back the 3' G-overhang enables it to invade distal, duplex regions of the telomere and base pair with the complementary C-rich strand, giving rise to the t-loop structure. See text for details.

In vertebrates stabilization of telomeres and formation of the t-loop are facilitated by the shelterin complex, which consists of the following six proteins: Telomeric Repeat-binding Factors 1 and 2 (TRF1 and TRF2), repressor and activator protein 1 (RAP1), TRF1-interacting nuclear protein 2 (TIN2), protection of telomeres 1 (POT1), and TIN2- and POT1-interacting protein (TPP1) [141]. Whereas TRF1 and TRF2 bind the double-stranded region of the telomere, POT1 has affinity specifically for the single-stranded G-overhang (Fig. 3). TIN2 bridges TRF1 and TRF2, while TPP1 bridges TIN2 and POT1. RAP1 is recruited to telomeres via its interaction with TRF2 [140]. Details regarding the vertebrate shelterin complex, its homologs in other organisms, and other facets of telomere structure and function are examined elsewhere in this issue.

The importance of the MRX complex for normal telomere maintenance was first recognized many years ago when deletion or disruption of *S. cerevisiae* Rad50 [145], Mre11, or Xrs2 [146] were reported to result in shortened telomeres and cell senescence [147]. Lundblad and co-workers subsequently demonstrated MRX to be in the same *S. cerevisiae* epistasis group as telomerase [148]. Consistent with this notion, *S. cerevisiae* MRX associates with telomeres in late S-phase [149] when yeast telomeres are synthesized, and in the absence of functional MRX the single-stranded telomeric DNA binding protein Cdc13 is unable to bind to telomeres [149] and [150]. Multiple groups have independently demonstrated that in both *S. cerevisiae* and *S. pombe* telomere length [151] and the specific formation of G-overhangs [149] and [152] are unaffected by the nuclease-inactivating D56N or H125N mutations of Mre11. Moreover, telomerase-mediated replication of telomeres is defective in Mre11 null cells but not in the Mre11-D56N or Mre11-H125N backgrounds [153]. Targeting *S. cerevisiae* telomerase to telomeres via fusion with Cdc13 overcomes telomere maintenance defects resulting from a non-functional MRX complex [153]. Collectively, these observations suggest that MRX facilitates the recruitment of telomerase to telomeres. Consistent with this, in *S. cerevisiae* the specific recruitment of telomerase to telomeres during the S phase is abolished by the absence of Mre11 [149]. While the MRX-mediated recruitment of yeast telomerase to telomeres could involve a direct interaction between these proteins, it could also simply be a consequence of MRX promoting the processing of telomeres into a form suitable for telomerase sequestration. Considering the well established role of MRX as a promoter of $5' \rightarrow 3'$ resection at DSBs, the specificity of Cdc13 for ssDNA, and the affinity of Cdc13 for telomerase, it is tempting to speculate that MRX works with Sae2 to generate short 3' overhangs that are bound by Cdc13 and subsequently recruit telomerase.

The first indication that MRN may also contribute to telomere maintenance in higher eukaryotes was the identification of Nbs1 and Mre11 sequestered at telomeres in meiotic human fibroblasts [154]. It was subsequently shown that MRN specifically associates with the TRF2 component of shelterin, and that Nbs1 accumulates at telomeres in S phase but not during G1 or G2 [155]. Since TRF2 does not associate with ionizing radiation-induced DSBs, this interaction appears to occur exclusively within the telomere microenvironment.

Cultured human NBS fibroblasts display shortened telomeres. In these cells, the coexpression of Nbs1 along with telomerase results in longer telomeres than the expression of telomerase alone [156]. This suggests that, analogous to the situation in yeast, human MRN facilitates telomerase activity at telomeres. Further supporting this notion, knockdown of Mre11 or Nbs1 in cultured human cancer cells specifically reduces the length of 3' G-overhangs, but has no effect in cells that do not express telomerase [157]. Collectively, these data demonstrate that MRN facilitates the action of telomerase at chromosome termini in mammalian cells. Whether MRN facilitates telomerase activity by modifying telomere ends, opening up the t-loop, altering chromatin structure, or by directly associating with telomerase remains to be determined. In telomerase-negative primary human cells, chromatin immunoprecipitation revealed that Mre11, phosphorylated Nbs1, and ATM were bound to telomeres in the G2 phase of the cell cycle [158]. This study suggested that telomeres become accessible in G2 and are recognized as DNA damage. A localized DNA damage response at telomeres may therefore be required for recruiting the processing machinery that is responsible for formation of the end protection complex.

Telomeres can be rendered dysfunctional by removal of the telomere DNA binding protein TRF2 from the shelterin complex. These "uncapped" telomeres are recognized as DSBs, and result in ATM activation, phosphorylation of Chk2 and H2AX, the formation of 53BP1-associated telomere-induced DNA damage foci (TIF), and NHEJ-mediated chromosome fusion [159]. The role of MRN at dysfunctional telomeres has recently been studied in embryonic fibroblasts derived from mice harboring MRN mutations and deletions, combined with TRF2 removal by conditional deletion or shRNA knockdown [160], [161] and [162]. These studies demonstrate that MRN is required for ATM signaling in response to telomere dysfunction. When TRF2 levels are depleted by shRNA, ATM activation and TIF formation are reduced in the Mre11$^{/}$ null background but remain robust in the nuclease deficient Mre11$^{H129N/}$ background [162]. This suggests that the MRN complex functions to detect and signal the presence of dysfunctional uncapped telomeres, and that this capability does not depend on the Mre11 nuclease activity.

Studies of dysfunctional telomeres performed in TRF2 deficient cells, have revealed a complex role for MRN in telomere fusion by NHEJ. Depending on the stage of the cell cycle and the specific structure of the telomere terminus, MRN can either promote or suppress the NHEJ-mediated fusion of dysfunctional telomeres. Conditional double-knockout of Nbs1 and TRF2 results in abrogated fusions in G1, due to defects in ATM-dependent signaling [161]. Although in G1 the MRN complex may promote NHEJ, the role of MRN after replication in G2 appears to be very different [161], and here the ability of MRN to promote end-to-end chromosome fusions at uncapped telomeres depends on Mre11's nuclease activity [162]. In another study employing TRF2 knockdown cells, the number of chromosome fusions in the Mre11$^{/}$ and Mre11$^{H129N/}$ backgrounds was 15-fold lower than that observed in an Mre11 active background [162]. To test whether Mre11 promotes NHEJ of TRF2-uncapped telomeres by removing the 3' G-overhang, an in-gel hybridization assay was employed. In contrast to cells with functional Mre11 where the 3' overhang is rapidly degraded, in Mre11$^{/}$ and Mre11$^{H129N/}$ cells the overhang persists [162]. This suggests that the Mre11 nuclease activity is required for processing of 3' overhangs to allow efficient NHEJ of telomere ends when rendered dysfunctional by TRF2 removal. An especially illuminating finding of these TRF2-uncapped telomere studies was that even though they occurred much less frequently, telomere–telomere fusions still occurred in the absence of functional MRN [160], [161] and [162]. Importantly, the majority (~90%) of these residual telomere fusions involved the leading strands of sister chromatids. In the Mre11$^{H129N/}$ background only ~60% of telomere–telomere fusions involved the leading strands of sister chromatids[162]. The authors suggested that this can be explained by the structural differences between telomere termini generated by leading versus lagging-strand replication. Leading-strand replication generates a blunt-ended telomere terminus that can readily be fused via NHEJ without prior nuclease processing. In contrast, lagging-strand replication generates a 3' telomeric overhang, which is incompatible with DNA ligation [159] and would therefore require nuclease processing prior to fusion. Thus, within a TRF2 deficient background the MRN complex appears to *prevent* the fusion of newly replicated leading strand telomeres by promoting 5' end resection to give NHEJ-incompatible 3' overhangs. It will be interesting to determine the extent to which CtIP and other end-processing factors are involved here.

The cellular response to dysfunctional telomeres is in many ways similar to the response induced by non-telomeric DSBs. In both of these contexts MRN is required for ATM activation, and many of the factors that accumulate at ionizing radiation-induced foci (IRIF) also accumulate at TIF. Moreover, both damage-induced DSBs and dysfunctional telomeres can lead to the same signaling pathways, cell cycle arrest, and apoptosis [163] and [164]. Differences between damage-induced DSBs and uncapped telomeres include (i) the fact that TRF2 suppresses ATM activation only within the telomere micro-environment since it is abundant at chromosome ends but not elsewhere in the nucleus [165], and (ii) DNA processing is not required for the ATM-mediated damage response at telomeres [159].

MRN AND EXOGENOUS DNA

In addition to the situation at the termini of linear eukaryotic chromosomes, the issues of end recognition and protection also emerge when exogenous extra-chromosomal DNA is encountered in the nucleus. For example, when the genome of a linear DNA virus is delivered to the nucleus of an infected cell it may be perceived as a DSB, and therefore has the potential to trigger the endogenous DNA damage response[166]. Some viruses have therefore evolved elaborate schemes to ensure that detrimental processing of their genomic termini does not take place during viral replication. In addition to its role as a sensor of cellular DNA ends, the MRN complex has emerged as a detector of viral genomes, with a central role in the cellular response to viral DNA [166]. This ability of the MRN complex to detect DNA ends can be either beneficial or detrimental to the virus lifecycle.

One of the first examples of a virus that interacts with the MRN complex came from the dsDNA virus Adenovirus. It has been suggested that one way the linear Adenovirus genome is protected is through virally-encoded proteins that induce the degradation and mislocalization of components of the cellular DNA damage machinery, including the MRN complex [167]. Mutants of Adenovirus unable to attenuate functions of the MRN complex are defective in viral replication and progeny production [168], [169],[170] and [171]. One phenotype of these mutant viruses is that the viral genome is ligated into concatamers through end-to-end joining in a process that requires both MRN and NHEJ factors [167] and [172]. During infection with Adenovirus mutants unable to inactivate the MRN complex, the cellular DNA damage signaling responses are also activated [69]. Attenuation of signaling and checkpoint activation by viral proteins that degrade or mislocalize MRN demonstrated upstream functions of this complex in both ATM and ATR signaling pathways [69] and [90]. The termini of the Adenovirus genome are also protected by a covalently-attached, virally-encoded terminal protein, which is important for initiation of viral DNA replication. We have suggested that removal of this protein from the end of the viral genome is analogous to removal of Spo11 from DSBs during meiotic recombination [167], and we propose that the combined action of the MRN complex together with CtIP is required for processing of the protein-blocked ends.

Beyond Adenovirus, the MRN complex has been implicated in the response to many different viruses[166]. It is found at viral replication centers during the early stages of SV40 infection, and components of the complex may be specifically downregulated late during infection in an ATM-dependent manner [173]. Nbs1 was shown to interact with the SV40 encoded T-antigen, leading to the discovery that Nbs1 can suppress rereplication of genomes during S phase [174]. The MRN complex is also found at virally-induced replication centers in the nuclei of cells infected by many different viruses. For example, MRN is present within the globular compartments formed during HSV-1 infection, and in this example Mre11 is beneficial to the early stages of viral replication [175]. There is also evidence that MRN may be downregulated late during HSV-1 infection, similar to the scenario with SV40 [176]. In addition to activation of signaling in response to viral genetic material, the MRN complex also plays a role in circularization, episomal maintenance and integration of viral genomes. These studies of the MRN complex in the context of virus infection have revealed its role as part of an anti-viral defense and have highlighted ways in which cellular repair pathways can be exploited by invading genomes [166].

CONCLUSIONS

The MRN complex sits at a central position in a complex network that senses, signals, and ultimately facilitates repair of DNA damage. It has crucial roles as a sensor of DSBs, and in activating the signal transduction cascades that lead to cell cycle checkpoints. It also plays pivotal roles in regulating repair pathway selection, as well as the actual DNA repair processes of both NHEJ and HR. A better understanding of the activities of this multifaceted complex will explain the requirements for maintaining genomic integrity, and the malignancies that arise in patients with genome instability disorders. This will also open up new opportunities to consider chemical ways to disrupt specific branches of the pathways controlled by MRN, and in this way sensitize tumor cells to DNA damaging cancer therapeutics. A forward chemical genetic screen has already identified an inhibitor of MRN that prevents MRN-dependent activation of ATM, the G2/M checkpoint, and homology-directed repair in mammalian cells [177].

The equilibrium between telomere function and dysfunction depends on a large number of factors, and the roles of MRN in this balance are only just beginning to be elucidated. While MRN can promote telomere function by recruiting telomerase to properly capped telomeres and by preventing the fusion of newly replicated leading strand telomeres, it can also exacerbate telomere dysfunction by degrading the G-overhang and thereby promoting telomere fusion. The dual functions of MRN in DNA processing and activation of damage signaling, are evident in its response to both genomic DSBs and telomeres. The importance of MRN to telomere metabolism has been demonstrated in a wide range of organisms, including *S. pombe*, *Kluyveromyces lactis*, *Arabidopsis thaliana*, *Drosophila melanogaster*, and *Homo sapiens*. Differences between these systems may be exploited to generate a detailed model of MRN function at telomeres, and will help us to navigate the complicated relationship between telomeres and the DNA damage machinery.

ACKNOWLEDGMENTS

We apologize to the many groups whose primary research papers could not be cited due to space constraints. We thank our colleagues in the field of DNA repair for helpful discussions and members of the Weitzman lab for comments on the manuscript. Work on the MRN complex in the Weitzman lab has been supported by grants from the National Institutes of Health (AI067952, CA097093 and AI051686) and a Pioneer Developmental Chair from the Salk Institute. B.J.L. is supported by a postdoctoral Ruth L. Kirschstein National Research Service Award (NIH/NCI T32 CA009523) and N.I.O. is supported in part by a gift from the H.A. & Mary K. Chapman Charitable Trust.

REFERENCES

1. , M., De Haro, L.P. and Nickoloff, J.A. (2008) Regulation of DNA double-strand break repair pathway choice. Cell Res. 18, 134–147.
2. Bernstein, K.A. and Rothstein, R. (2009) At loose ends: resecting a doublestrand break. Cell 137, 807–810.
3. Taylor, E.M., Cecillon, S.M., Bonis, A., Chapman, J.R., Povirk, L.F. and Lindsay, H.D. (2010) The Mre11/Rad50/Nbs1 complex functions in resection-based DNA end joining in Xenopus laevis. Nucleic Acids Res. 38, 441–454.
4. Rass, E., Grabarz, A., Plo, I., Gautier, J., Bertrand, P. and Lopez, B.S. (2009) Role of Mre11 in chromosomal nonhomologous end joining in mammalian cells. Nat. Struct. Mol. Biol. 16, 819–824.
5. Xie, A., Kwok, A. and Scully, R. (2009) Role of mammalian Mre11 in classical and alternative nonhomologous end joining. Nat. Struct. Mol. Biol. 16, 814–818.
6. Dinkelmann, M., Spehalski, E., Stoneham, T., Buis, J., Wu, Y., Sekiguchi, J.M. and Ferguson, D.O. (2009) Multiple functions of MRN in end-joining pathways during isotype class switching. Nat. Struct. Mol. Biol. 16, 808–813.
7. Zha, S., Boboila, C. and Alt, F.W. (2009) Mre11: roles in DNA repair beyond homologous recombination. Nat. Struct. Mol. Biol. 16, 798–800.
8. Ogawa, H., Johzuka, K., Nakagawa, T., Leem, S.H. and Hagihara, A.H. (1995) Functions of the yeast meiotic recombination genes, MRE11 and MRE2. Adv. Biophys. 31, 67–76.
9. Usui, T., Ohta, T., Oshiumi, H., Tomizawa, J., Ogawa, H. and Ogawa, T. (1998) Complex formation and functional versatility of Mre11 of budding yeast in recombination. Cell 95, 705–716.

10. Dolganov, G.M., Maser, R.S., Novikov, A., Tosto, L., Chong, S., Bressan, D.A. and Petrini, J.H. (1996) Human Rad50 is physically associated with human Mre11: identification of a conserved multiprotein complex implicated in recombinational DNA repair. Mol. Cell Biol. 16, 4832–4841.
11. Trujillo, K.M., Yuan, S.S., Lee, E.Y. and Sung, P. (1998) Nuclease activities in a complex of human recombination and DNA repair factors Rad50, Mre11, and p95. J. Biol. Chem. 273, 21447–21450.
12. Hopkins, B.B. and Paull, T.T. (2008) The P. furiosus mre11/rad50 complex promotes 5' strand resection at a DNA double-strand break. Cell 135, 250–260.
13. Williams, R.S., Williams, J.S. and Tainer, J.A. (2007) Mre11-Rad50-Nbs1 is a keystone complex connecting DNA repair machinery, double-strand break signaling, and the chromatin template. Biochem. Cell Biol. 85, 509–520.
14. D'Amours, D. and Jackson, S.P. (2002) The Mre11 complex: at the crossroads of DNA repair and checkpoint signalling. Nat. Rev. Mol. Cell Biol. 3, 317–327.
15. Williams, R.S. et al. (2008) Mre11 dimers coordinate DNA end bridging and nuclease processing in double-strand-break repair. Cell 135, 97–109.
16. de Jager, M., Dronkert, M.L., Modesti, M., Beerens, C.E., Kanaar, R. and van Gent, D.C. (2001) DNA-binding and strand-annealing activities of human Mre11: implications for its roles in DNA double-strand break repair pathways. Nucleic Acids Res. 29, 1317–1325.
17. Paull, T.T. and Gellert, M. (1999) Nbs1 potentiates ATP-driven DNA unwinding and endonuclease cleavage by the Mre11/Rad50 complex. Genes Dev. 13, 1276–1288.
18. Paull, T.T. and Gellert, M. (1998) The 30 to 50 exonuclease activity of Mre 11 facilitates repair of DNA double-strand breaks. Mol. Cell 1, 969–979.
19. Zhuang, J., Jiang, G., Willers, H. and Xia, F. (2009) Exonuclease function of human Mre11 promotes deletional

nonhomologous end joining. J. Biol. Chem. 284, 30565–30573.

20. Milman, N., Higuchi, E. and Smith, G.R. (2009) Meiotic DNA double-strand break repair requires two nucleases, MRN and Ctp1, to produce a single size class of Rec12 (Spo11)-oligonucleotide complexes. Mol. Cell Biol. 29, 5998–6005.

21. Farah, J.A., Cromie, G.A. and Smith, G.R. (2009) Ctp1 and exonuclease 1, alternative nucleases regulated by the MRN complex, are required for efficient meiotic recombination. Proc. Natl. Acad. Sci. USA 106, 9356–9361.

22. Mimitou, E.P. and Symington, L.S. (2009) DNA end resection: many nucleases make light work. DNA Repair (Amst) 8, 983–995.

23. van der Linden, E., Sanchez, H., Kinoshita, E., Kanaar, R. and Wyman, C. (2009) RAD50 and NBS1 form a stable complex functional in DNA binding and tethering. Nucleic Acids Res. 37, 1580–1588.

24. Stewart, G.S. et al. (1999) The DNA double-strand break repair gene hMRE11 is mutated in individuals with an ataxia-telangiectasia-like disorder. Cell 99, 577–587.

25. Zhong, H., Bryson, A., Eckersdorff, M. and Ferguson, D.O. (2005) Rad50 depletion impacts upon ATR-dependent DNA damage responses. Hum. Mol. Genet. 14, 2685–2693.

26. Hopfner, K.P., Karcher, A., Shin, D.S., Craig, L., Arthur, L.M., Carney, J.P. and Tainer, J.A. (2000) Structural biology of Rad50 ATPase: ATP-driven conformational control in DNA double-strand break repair and the ABCATPase superfamily. Cell 101, 789–800.

27. Hopfner, K.P., Karcher, A., Craig, L., Woo, T.T., Carney, J.P. and Tainer, J.A. (2001) Structural biochemistry and interaction architecture of the DNA double-strand break repair Mre11 nuclease and Rad50-ATPase. Cell 105, 473–485.

28. de Jager, M., van Noort, J., van Gent, D.C., Dekker, C., Kanaar, R. and Wyman, C. (2001) Human Rad50/Mre11 is a flexible complex that can tether DNA ends. Mol. Cell 8, 1129–1135.

29. Hopfner, K.P. et al. (2002) The Rad50 zinc-hook is a structure joining Mre11 complexes in DNA recombination and repair. Nature 418, 562–566.
30. Cahill, D. and Carney, J.P. (2007) Dimerization of the Rad50 protein is independent of the conserved hook domain. Mutagenesis 22, 269–274.
31. Wiltzius, J.J., Hohl, M., Fleming, J.C. and Petrini, J.H. (2005) The Rad50 hook domain is a critical determinant of Mre11 complex functions. Nat. Struct. Mol. Biol. 12, 403–407.
32. Lloyd, J. et al. (2009) A supramodular FHA/BRCT-repeat architecture mediates Nbs1 adaptor function in response to DNA damage. Cell 139, 100–111.
33. Williams, R.S. et al. (2009) Nbs1 flexibly tethers Ctp1 and Mre11-Rad50 to coordinate DNA double-strand break processing and repair. Cell 139, 87–99.
34. Desai-Mehta, A., Cerosaletti, K.M. and Concannon, P. (2001) Distinct functional domains of nibrin mediate Mre11 binding, focus formation, and nuclear localization. Mol. Cell Biol. 21, 2184–2191.
35. Falck, J., Coates, J. and Jackson, S.P. (2005) Conserved modes of recruitment of ATM, ATR and DNA-PKcs to sites of DNA damage. Nature 434, 605–611.
36. You, Z., Chahwan, C., Bailis, J., Hunter, T. and Russell, P. (2005) ATM activation and its recruitment to damaged DNA require binding to the C terminus of Nbs1. Mol. Cell Biol. 25, 5363–5379.
37. Stracker, T.H., Morales, M., Couto, S.S., Hussein, H. and Petrini, J.H.J. (2007) The carboxy terminus of NBS1 is required for induction of apoptosis by the MRE11 complex. Nature 447, 218–221.
38. Carney, J.P. et al. (1998) The hMre11/hRad50 protein complex and Nijmegen breakage syndrome: linkage of double-strand break repair to the cellular DNA damage response. Cell 93, 477–486.

39. Varon, R. et al. (1998) Nibrin, a novel DNA double-strand break repair protein, is mutated in Nijmegen breakage syndrome. Cell 93, 467–476.
40. Antoccia, A., Kobayashi, J., Tauchi, H., Matsuura, S. and Komatsu, K. (2006) Nijmegen breakage syndrome and functions of the responsible protein, NBS1. Genome Dyn. 1, 191–205.
41. Taylor, A.M., Groom, A. and Byrd, P.J. (2004) Ataxia-telangiectasia-like disorder (ATLD)-its clinical presentation and molecular basis. DNA Repair (Amst) 3, 1219–1225.
42. Waltes, R. et al. (2009) Human RAD50 deficiency in a Nijmegen breakage syndrome-like disorder. Am. J. Hum. Genet. 84, 605–616.
43. Dzikiewicz-Krawczyk, A. (2008) The importance of making ends meet: mutations in genes and altered expression of proteins of the MRN complex and cancer. Mutat. Res. 659, 262–273.
44. Theunissen, J.-W.F., Kaplan, M.I., Hunt, P.A., Williams, B.R., Ferguson, D.O., Alt, F.W. and Petrini, J.H.J. (2003) Checkpoint failure and chromosomal instability without lymphomagenesis in Mre11(ATLD1/ATLD1) mice. Mol. Cell 12, 1511–1523.
45. Kang, J., Bronson, R.T. and Xu, Y. (2002) Targeted disruption of NBS1 reveals its roles in mouse development and DNA repair. EMBO J. 21, 1447–1455.
46. Difilippantonio, S. et al. (2005) Role of Nbs1 in the activation of the Atm kinase revealed in humanized mouse models. Nat. Cell Biol. 7, 675–685.
47. Difilippantonio, S. et al. (2007) Distinct domains in Nbs1 regulate irradiationinduced checkpoints and apoptosis. J. Exp. Med. 204, 1003–1011.
48. Bender, C.F. et al. (2002) Cancer predisposition and hematopoietic failure in Rad50(S/S) mice. Genes Dev. 16, 2237–2251.
49. Kracker, S. et al. (2005) Nibrin functions in Ig class-switch recombination. Proc. Natl. Acad. Sci. USA 102, 1584–1589.

50. Frappart, P.-O., Tong, W.-M., Demuth, I., Radovanovic, I., Herceg, Z., Aguzzi, A., Digweed, M. and Wang, Z.-Q. (2005) An essential function for NBS1 in the prevention of ataxia and cerebellar defects. Nat. Med. 11, 538–544.
51. Reina-San-Martin, B., Nussenzweig, M.C., Nussenzweig, A. and Difilippantonio, S. (2005) Genomic instability, endoreduplication, and diminished Ig class-switch recombination in B cells lacking Nbs1. Proc. Natl. Acad. Sci. USA 102, 1590–1595.
52. Williams, B.R., Mirzoeva, O.K., Morgan, W.F., Lin, J., Dunnick, W. and Petrini, J.H.J. (2002) A murine model of Nijmegen breakage syndrome. Curr. Biol. 12, 648–653.
53. [53] Luo, G., Yao, M.S., Bender, C.F., Mills, M., Bladl, A.R., Bradley, A. and Petrini, J.H. (1999) Disruption of mRad50 causes embryonic stem cell lethality, abnormal embryonic development, and sensitivity to ionizing radiation. Proc. Natl. Acad. Sci. USA 96, 7376–7381.
54. Cherry, S.M., Adelman, C.A., Theunissen, J.W., Hassold, T.J., Hunt, P.A. and Petrini, J.H. (2007) The Mre11 complex influences DNA repair, synapsis, and crossing over in murine meiosis. Curr. Biol. 17, 373–378.
55. Frappart, P.O. and McKinnon, P.J. (2006) Ataxia-telangiectasia and related diseases. Neuromol. Med. 8, 495–511.
56. Baranes, K. et al. (2009) Conditional inactivation of the NBS1 gene in the mouse central nervous system leads to neurodegeneration and disorganization of the visual system. Exp. Neurol. 218, 24–32.
57. Shull, E.R.P., Lee, Y., Nakane, H., Stracker, T.H., Zhao, J., Russell, H.R., Petrini, J.H.J. and McKinnon, P.J. (2009) Differential DNA damage signaling accounts for distinct neural apoptotic responses in ATLD and NBS. Genes Dev. 23, 171–180.
58. Lavin, M.F. (2007) ATM and the Mre11 complex combine to recognize and signal DNA double-strand breaks. Oncogene 26, 7749–7758.

59. Borde, V. and Cobb, J. (2009) Double functions for the Mre11 complex during DNA double-strand break repair and replication. Int. J. Biochem. Cell Biol. 41, 1249–1253.
60. Borde, V. (2007) The multiple roles of the Mre11 complex for meiotic recombination. Chromosome Res. 15, 551–563.
61. Jackson, S.P. and Bartek, J. (2009) The DNA-damage response in human biology and disease. Nature 461, 1071–1078.
62. Harper, J.W. and Elledge, S.J. (2007) The DNA damage response: ten years after. Mol. Cell 28, 739–745.
63. Rogakou, E.P., Pilch, D.R., Orr, A.H., Ivanova, V.S. and Bonner, W.M. (1998) DNA double-stranded breaks induce histone H2AX phosphorylation on serine 139. J. Biol. Chem. 273, 5858–5868.
64. Dickey, J.S., Redon, C.E., Nakamura, A.J., Baird, B.J., Sedelnikova, O.A. and Bonner, W.M. (2009) H2AX: functional roles and potential applications. Chromosoma 118, 683–692.
65. Schleker, T., Nagai, S. and Gasser, S.M. (2009) Posttranslational modifications of repair factors and histones in the cellular response to stalled replication forks. DNA Repair (Amst) 8, 1089–1100.
66. Huen, M.S. and Chen, J. (2008) The DNA damage response pathways: at the crossroad of protein modifications. Cell Res. 18, 8–16.
67. Matsuoka, S. et al. (2007) ATM and ATR substrate analysis reveals extensive protein networks responsive to DNA damage. Science 316, 1160–1166.
68. Petrini, J.H. and Stracker, T.H. (2003) The cellular response to DNA doublestrand breaks: defining the sensors and mediators. Trends Cell Biol. 13, 458–462.
69. [69] Carson, C.T., Schwartz, R.A., Stracker, T.H., Lilley, C.E., Lee, D.V. and Weitzman, M.D. (2003) The Mre11 complex is required for ATM activation and the G2/M checkpoint. EMBO J. 22, 6610–6620.

70. Girard, P.M., Riballo, E., Begg, A.C., Waugh, A. and Jeggo, P.A. (2002) Nbs1 promotes ATM dependent phosphorylation events including those required for G1/S arrest. Oncogene 21, 4191–4199.
71. Uziel, T., Lerenthal, Y., Moyal, L., Andegeko, Y., Mittelman, L. and Shiloh, Y. (2003) Requirement of the MRN complex for ATM activation by DNA damage. EMBO J. 22, 5612–5621.
72. Maser, R.S., Monsen, K.J., Nelms, B.E. and Petrini, J.H. (1997) HMre11 and hRad50 nuclear foci are induced during the normal cellular response to DNA double-strand breaks. Mol. Cell Biol. 17, 6087–6096.
73. Mirzoeva, O.K. and Petrini, J.H. (2001) DNA damage-dependent nuclear dynamics of the Mre11 complex. Mol. Cell Biol. 21, 281–288.
74. Nelms, B.E., Maser, R.S., MacKay, J.F., Lagally, M.G. and Petrini, J.H. (1998) In situ visualization of DNA double-strand break repair in human fibroblasts. Science 280, 590–592.
75. Paull, T.T., Rogakou, E.P., Yamazaki, V., Kirchgessner, C.U., Gellert, M. and Bonner, W.M. (2000) A critical role for histone H2AX in recruitment of repair factors to nuclear foci after DNA damage. Curr. Biol. 10, 886–895.
76. Lukas, C., Falck, J., Bartkova, J., Bartek, J. and Lukas, J. (2003) Distinct spatiotemporal dynamics of mammalian checkpoint regulators induced by DNA damage. Nat. Cell Biol. 5, 255–260.
77. Bakkenist, C.J. and Kastan, M.B. (2003) DNA damage activates ATM through intermolecular autophosphorylation and dimer dissociation. Nature 421, 499–506.
78. Lee, J.H., Goodarzi, A.A., Jeggo, P.A. and Paull, T.T. (2010) 53BP1 promotes ATM activity through direct interactions with the MRN complex. EMBO J. 29, 574–585.
79. So, S., Davis, A.J. and Chen, D.J. (2009) Autophosphorylation at serine 1981 stabilizes ATM at DNA damage sites. J. Cell Biol. 187, 977–990.

80. Lee, J.H. and Paull, T.T. (2005) ATM activation by DNA double-strand breaks through the Mre11–Rad50–Nbs1 complex. Science 308, 551–554.
81. Berkovich, E., Monnat Jr., R.J. and Kastan, M.B. (2007) Roles of ATM and NBS1 in chromatin structure modulation and DNA double-strand break repair. Nat. Cell Biol. 9, 683–690.
82. Kitagawa, R., Bakkenist, C.J., McKinnon, P.J. and Kastan, M.B. (2004) Phosphorylation of SMC1 is a critical downstream event in the ATM–NBS1–BRCA1 pathway. Genes Dev. 18, 1423–1438.
83. Lee, J.S. (2007) Activation of ATM-dependent DNA damage signal pathway by a histone deacetylase inhibitor, trichostatin A. Cancer Res. Treat. 39, 125–130.
84. Jang, E.R., Choi, J.D., Park, M.A., Jeong, G., Cho, H. and Lee, J.S. (2010) ATM modulates transcription in response to histone deacetylase inhibition as part of its DNA damage response. Exp. Mol. Med. 42, 195–204.
85. Sun, Y., Jiang, X. and Price, B.D. (2010) Tip60: connecting chromatin to DNA damage signaling. Cell Cycle 9, 930–936.
86. Kanu, N. and Behrens, A. (2008) ATMINistrating ATM signalling: regulation of ATM by ATMIN. Cell Cycle 7, 3483–3486.
87. Jazayeri, A., Falck, J., Lukas, C., Bartek, J., Smith, G.C., Lukas, J. and Jackson, S.P. (2006) ATM- and cell cycle-dependent regulation of ATR in response to DNA double-strand breaks. Nat. Cell Biol. 8, 37–45.
88. Shiotani, B. and Zou, L. (2009) Single-stranded DNA orchestrates an ATM-toATR switch at DNA breaks. Mol. Cell 33, 547–558.
89. Buis, J. et al. (2008) Mre11 nuclease activity has essential roles in DNA repair and genomic stability distinct from ATM activation. Cell 135, 85–96.
90. Carson, C.T. et al. (2009) Mislocalization of the MRN complex prevents ATR signaling during adenovirus infection. EMBO J. 28, 652–662.

91. Manthey, K.C., Opiyo, S., Glanzer, J.G., Dimitrova, D., Elliott, J. and Oakley, G.G. (2007) NBS1 mediates ATR-dependent RPA hyperphosphorylation following replication-fork stall and collapse. J. Cell Sci. 120, 4221–4229.
92. Olson, E., Nievera, C.J., Lee, A.Y.-L., Chen, L. and Wu, X. (2007) The Mre11–Rad50–Nbs1 complex acts both upstream and downstream of ataxia telangiectasia mutated and Rad3-related protein (ATR) to regulate the Sphase checkpoint following UV treatment. J. Biol. Chem. 282, 22939–22952.
93. Stiff, T., Reis, C., Alderton, G.K., Woodbine, L., O'Driscoll, M. and Jeggo, P.A. (2005) Nbs1 is required for ATR-dependent phosphorylation events. EMBO J. 24, 199–208.
94. Jazayeri, A., Balestrini, A., Garner, E., Haber, J.E. and Costanzo, V. (2008) Mre11–Rad50–Nbs1-dependent processing of DNA breaks generates oligonucleotides that stimulate ATM activity. EMBO J. 27, 1953–1962.
95. Reinhardt, H.C. and Yaffe, M.B. (2009) Kinases that control the cell cycle in response to DNA damage: Chk1, Chk2, and MK2. Curr. Opin. Cell Biol. 21, 245–255.
96. Tsukuda, T., Fleming, A.B., Nickoloff, J.A. and Osley, M.A. (2005) Chromatin remodelling at a DNA double-strand break site in Saccharomyces cerevisiae. Nature 438, 379–383.
97. Sun, Y., Jiang, X., Xu, Y., Ayrapetov, M.K., Moreau, L.A., Whetstine, J.R. and Price, B.D. (2009) Histone H3 methylation links DNA damage detection to activation of the tumour suppressor Tip60. Nat. Cell Biol. 11, 1376–1382.
98. Conde, F., Refolio, E., Cordon-Preciado, V., Cortes-Ledesma, F., Aragon, L., Aguilera, A. and San-Segundo, P.A. (2009) The Dot1 histone methyltransferase and the Rad9 checkpoint adaptor contribute to cohesindependent double-strand break repair by sister chromatid recombination in Saccharomyces cerevisiae. Genetics 182, 437–446.
99. Galanty, Y., Belotserkovskaya, R., Coates, J., Polo, S., Miller, K.M. and Jackson, S.P. (2009) Mammalian SUMO E3-ligases

PIAS1 and PIAS4 promote responses to DNA double-strand breaks. Nature 462, 935–939.

100. Messick, T.E. and Greenberg, R.A. (2009) The ubiquitin landscape at DNA double-strand breaks. J. Cell Biol. 187, 319–326.
101. Mao, Z., Bozzella, M., Seluanov, A. and Gorbunova, V. (2008) Comparison of nonhomologous end joining and homologous recombination in human cells. DNA Repair (Amst) 7, 1765–1771.
102. Aguilera, A. and Gomez-Gonzalez, B. (2008) Genome instability: a mechanistic view of its causes and consequences. Nat. Rev. Genet. 9, 204–217.
103. Branzei, D. and Foiani, M. (2008) Regulation of DNA repair throughout the cell cycle. Nat. Rev. Mol. Cell Biol. 9, 297–308.
104. Hartlerode, A.J. and Scully, R. (2009) Mechanisms of double-strand break repair in somatic mammalian cells. Biochem. J. 423, 157–168.
105. Ding, D.Q., Haraguchi, T. and Hiraoka, Y. (2010) From meiosis to postmeiotic events: alignment and recognition of homologous chromosomes in meiosis. FEBS J. 277, 565–570.
106. Huertas, P. (2010) DNA resection in eukaryotes: deciding how to fix the break. Nat. Struct. Mol. Biol. 17, 11–16.
107. Aylon, Y., Liefshitz, B. and Kupiec, M. (2004) The CDK regulates repair of double-strand breaks by homologous recombination during the cell cycle. EMBO J. 23, 4868–4875.
108. Ira, G. et al. (2004) DNA end resection, homologous recombination and DNA damage checkpoint activation require CDK1. Nature 431, 1011–1017.
109. Dynan, W.S. and Yoo, S. (1998) Interaction of Ku protein and DNA-dependent protein kinase catalytic subunit with nucleic acids. Nucleic Acids Res. 26, 1551–1559.
110. Zhang, Y., Shim, E.Y., Davis, M. and Lee, S.E. (2009) Regulation of repair choice. Cdk1 suppresses recruitment of end joining

factors at DNA breaks. DNA Repair (Amst) 8, 1235–1241.
111. Clerici, M., Mantiero, D., Lucchini, G. and Longhese, M.P. (2005) The Saccharomyces cerevisiae Sae2 protein promotes resection and bridging of double strand break ends. J. Biol. Chem. 280, 38631–38638.
112. Lengsfeld, B.M., Rattray, A.J., Bhaskara, V., Ghirlando, R. and Paull, T.T. (2007) Sae2 is an endonuclease that processes hairpin DNA cooperatively with the Mre11/Rad50/Xrs2 complex. Mol. Cell 28, 638–651.
113. Neale, M.J., Pan, J. and Keeney, S. (2005) Endonucleolytic processing of covalent protein-linked DNA double-strand breaks. Nature 436, 1053–1057.
114. Limbo, O., Chahwan, C., Yamada, Y., de Bruin, R.A., Wittenberg, C. and Russell, P. (2007) Ctp1 is a cell-cycle-regulated protein that functions with Mre11 complex to control double-strand break repair by homologous recombination. Mol. Cell 28, 134–146.
115. Huertas, P. and Jackson, S.P. (2009) Human CtIP mediates cell cycle control of DNA end resection and double strand break repair. J. Biol. Chem. 284, 9558–9565.
116. Sartori, A.A. et al. (2007) Human CtIP promotes DNA end resection. Nature 450, 509–514.
117. You, Z. and Bailis, J.M. (2010) DNA damage and decisions: CtIP coordinates DNA repair and cell cycle checkpoints. Trends Cell Biol. 20, 402–409.
118. You, Z. et al. (2009) CtIP links DNA double-strand break sensing to resection. Mol. Cell 36, 954–969.
119. Mimitou, E.P. and Symington, L.S. (2008) Sae2, Exo1 and Sgs1 collaborate in DNA double-strand break processing. Nature 455, 770–774.
120. Zhu, Z., Chung, W.H., Shim, E.Y., Lee, S.E. and Ira, G. (2008) Sgs1 helicase and two nucleases Dna2 and Exo1 resect DNA double-strand break ends. Cell 134, 981–994.

121. Gravel, S., Chapman, J.R., Magill, C. and Jackson, S.P. (2008) DNA helicases Sgs1 and BLM promote DNA double-strand break resection. Genes Dev. 22, 2767–2772.
122. Yu, X. and Baer, R. (2000) Nuclear localization and cell cycle-specific expression of CtIP, a protein that associates with the BRCA1 tumor suppressor. J. Biol. Chem. 275, 18541–18549.
123. Yun, M.H. and Hiom, K. (2009) CtIP-BRCA1 modulates the choice of DNA double-strand-break repair pathway throughout the cell cycle. Nature 459, 460–463.
124. Dodson, G.E., Limbo, O., Nieto, D. and Russell, P. (2010) Phosphorylationregulated binding of Ctp1 to Nbs1 is critical for repair of DNA double-strand breaks. Cell Cycle 9, 1516–1522.
125. Huertas, P., Cortes-Ledesma, F., Sartori, A.A., Aguilera, A. and Jackson, S.P. (2008) CDK targets Sae2 to control DNA-end resection and homologous recombination. Nature 455, 689–692.
126. Kim, J.S., Krasieva, T.B., Kurumizaka, H., Chen, D.J., Taylor, A.M. and Yokomori, K. (2005) Independent and sequential recruitment of NHEJ and HR factors to DNA damage sites in mammalian cells. J. Cell Biol. 170, 341–347.
127. Wasko, B.M., Holland, C.L., Resnick, M.A. and Lewis, L.K. (2009) Inhibition of DNA double-strand break repair by the Ku heterodimer in mrx mutants of Saccharomyces cerevisiae. DNA Repair (Amst) 8, 162–169.
128. Fukushima, T. et al. (2001) Genetic analysis of the DNA-dependent protein kinase reveals an inhibitory role of Ku in late S-G2 phase DNA double-strand break repair. J. Biol. Chem. 276, 44413–44418.
129. Clikeman, J.A., Khalsa, G.J., Barton, S.L. and Nickoloff, J.A. (2001) Homologous recombinational repair of double-strand breaks in yeast is enhanced by MAT heterozygosity through yKU-dependent and -independent mechanisms. Genetics 157, 579–589.

130. Zhang, Y. et al. (2007) Role of Dnl4-Lif1 in nonhomologous end-joining repair complex assembly and suppression of homologous recombination. Nat. Struct. Mol. Biol. 14, 639–646.
131. Pierce, A.J., Hu, P., Han, M., Ellis, N. and Jasin, M. (2001) Ku DNA end-binding protein modulates homologous repair of double-strand breaks in mammalian cells. Genes Dev. 15, 3237–3242.
132. Wu, D., Topper, L.M. and Wilson, T.E. (2008) Recruitment and dissociation of nonhomologous end joining proteins at a DNA double-strand break in Saccharomyces cerevisiae. Genetics 178, 1237–1249.
133. Huang, J. and Dynan, W.S. (2002) Reconstitution of the mammalian DNA double-strand break end-joining reaction reveals a requirement for an Mre11/Rad50/NBS1-containing fraction. Nucleic Acids Res. 30, 667–674.
134. Di Virgilio, M. and Gautier, J. (2005) Repair of double-strand breaks by nonhomologous end joining in the absence of Mre11. J. Cell Biol. 171, 765–771.
135. Weterings, E. and Chen, D.J. (2008) The endless tale of nonhomologous endjoining. Cell Res. 18, 114–124.
136. Meek, K., Gupta, S., Ramsden, D.A. and Lees-Miller, S.P. (2004) The DNAdependent protein kinase: the director at the end. Immunol. Rev. 200, 132–141.
137. Gu, J. et al. (2010) DNA-PKcs regulates a single-stranded DNA endonuclease activity of Artemis. DNA Repair (Amst) 9, 429–437.
138. Yannone, S.M., Khan, I.S., Zhou, R.Z., Zhou, T., Valerie, K. and Povirk, L.F. (2008) Coordinate 5′ and 3′ endonucleolytic trimming of terminally blocked blunt DNA double-strand break ends by Artemis nuclease and DNAdependent protein kinase. Nucleic Acids Res. 36, 3354–3365.
139. Zhang, X. and Paull, T.T. (2005) The Mre11/Rad50/Xrs2 complex and nonhomologous end-joining of incompatible ends in S. cerevisiae. DNA Repair (Amst) 4, 1281–1294.

140. O'Sullivan, R.J. and Karlseder, J. (2010) Telomeres: protecting chromosomes against genome instability. Nat. Rev. Mol. Cell Biol. 11, 171–181.
141. Palm, W. and de Lange, T. (2008) How shelterin protects mammalian telomeres. Annu. Rev. Genet. 42, 301–334.
142. Denchi, E.L. (2009) Give me a break: how telomeres suppress the DNA damage response. DNA Repair (Amst) 8, 1118–1126.
143. Larrivee, M., LeBel, C. and Wellinger, R.J. (2004) The generation of proper constitutive G-tails on yeast telomeres is dependent on the MRX complex. Genes Dev. 18, 1391–1396.
144. Lydall, D. (2003) Hiding at the ends of yeast chromosomes: telomeres, nucleases and checkpoint pathways. J. Cell Sci. 116, 4057–4065.
145. Kironmai, K.M. and Muniyappa, K. (1997) Alteration of telomeric sequences and senescence caused by mutations in RAD50 of Saccharomyces cerevisiae. Genes Cells 2, 443–455.
146. Boulton, S.J. and Jackson, S.P. (1998) Components of the Ku-dependent nonhomologous end-joining pathway are involved in telomeric length maintenance and telomeric silencing. EMBO J. 17, 1819–1828.
147. Chamankhah, M. and Xiao, W. (1999) Formation of the yeast Mre11–Rad50– Xrs2 complex is correlated with DNA repair and telomere maintenance. Nucleic Acids Res. 27, 2072–2079.
148. Nugent, C.I., Bosco, G., Ross, L.O., Evans, S.K., Salinger, A.P., Moore, J.K., Haber, J.E. and Lundblad, V. (1998) Telomere maintenance is dependent on activities required for end repair of double-strand breaks. Curr. Biol. 8, 657–660.
149. Takata, H., Tanaka, Y. and Matsuura, A. (2005) Late S phase-specific recruitment of Mre11 complex triggers hierarchical assembly of telomere replication proteins in Saccharomyces cerevisiae. Mol. Cell 17, 573–583.

150. Diede, S.J. and Gottschling, D.E. (2001) Exonuclease activity is required for sequence addition and Cdc13p loading at a de novo telomere. Curr. Biol. 11, 1336–1340.
151. Moreau, S., Ferguson, J.R. and Symington, L.S. (1999) The nuclease activity of Mre11 is required for meiosis but not for mating type switching, end joining, or telomere maintenance. Mol. Cell Biol. 19, 556–566.
152. Tomita, K. et al. (2003) Competition between the Rad50 complex and the Ku heterodimer reveals a role for Exo1 in processing double-strand breaks but not telomeres. Mol. Cell Biol. 23, 5186–5197.
153. Tsukamoto, Y., Taggart, A.K. and Zakian, V.A. (2001) The role of the Mre11– Rad50–Xrs2 complex in telomerase-mediated lengthening of Saccharomyces cerevisiae telomeres. Curr. Biol. 11, 1328–1335.
154. Lombard, D.B. and Guarente, L. (2000) Nijmegen breakage syndrome disease protein and MRE11 at PML nuclear bodies and meiotic telomeres. Cancer Res. 60, 2331–2334.
155. Zhu, X.D., Kuster, B., Mann, M., Petrini, J.H. and de Lange, T. (2000) Cell cycleregulated association of RAD50/MRE11/NBS1 with TRF2 and human telomeres. Nat. Genet. 25, 347–352.
156. Ranganathan, V. et al. (2001) Rescue of a telomere length defect of Nijmegen breakage syndrome cells requires NBS and telomerase catalytic subunit. Curr. Biol. 11, 962–966.
157. Chai, W., Sfeir, A.J., Hoshiyama, H., Shay, J.W. and Wright, W.E. (2006) The involvement of the Mre11/Rad50/Nbs1 complex in the generation of Goverhangs at human telomeres. EMBO Rep. 7, 225–230.
158. Verdun, R.E., Crabbe, L., Haggblom, C. and Karlseder, J. (2005) Functional human telomeres are recognized as DNA damage in G2 of the cell cycle. Mol. Cell 20, 551–561.
159. Celli, G.B. and de Lange, T. (2005) DNA processing is not required for ATMmediated telomere damage response after TRF2 deletion. Nat. Cell Biol. 7, 712–718.

160. Attwooll, C.L., Akpinar, M. and Petrini, J.H. (2009) The mre11 complex and the response to dysfunctional telomeres. Mol. Cell Biol. 29, 5540–5551.

161. Dimitrova, N. and de Lange, T. (2009) Cell cycle-dependent role of MRN at dysfunctional telomeres: ATM signaling-dependent induction of nonhomologous end joining (NHEJ) in G1 and resection-mediated inhibition of NHEJ in G2. Mol. Cell Biol. 29, 5552–5563.

162. Deng, Y., Guo, X., Ferguson, D.O. and Chang, S. (2009) Multiple roles for MRE11 at uncapped telomeres. Nature 460, 914–918.

163. Takai, H., Smogorzewska, A. and de Lange, T. (2003) DNA damage foci at dysfunctional telomeres. Curr. Biol. 13, 1549–1556.

164. Karlseder, J., Broccoli, D., Dai, Y., Hardy, S. and de Lange, T. (1999) P53- and ATM-dependent apoptosis induced by telomeres lacking TRF2. Science 283, 1321–1325.

165. Karlseder, J., Hoke, K., Mirzoeva, O.K., Bakkenist, C., Kastan, M.B., Petrini, J.H. and de Lange, T. (2004) The telomeric protein TRF2 binds the ATM kinase and can inhibit the ATM-dependent DNA damage response. PLoS Biol. 2, E240.

166. Lilley, C.E., Schwartz, R.A. and Weitzman, M.D. (2007) Using or abusing: viruses and the cellular DNA damage response. Trends Microbiol. 15, 119–126.

167. Stracker, T.H., Carson, C.T. and Weitzman, M.D. (2002) Adenovirus oncoproteins inactivate the Mre11–Rad50–NBS1 DNA repair complex. Nature 418, 348–352.

168. Evans, J.D. and Hearing, P. (2005) Relocalization of the Mre11–Rad50–Nbs1 complex by the adenovirus E4 ORF3 protein is required for viral replication. J. Virol. 79, 6207–6215.

169. Mathew, S.S. and Bridge, E. (2007) The cellular Mre11 protein interferes with adenovirus E4 mutant DNA replication. Virology 365, 346–355.

170. Mathew, S.S. and Bridge, E. (2008) Nbs1-dependent binding of Mre11 to adenovirus E4 mutant viral DNA is important for inhibiting DNA replication. Virology 374, 11–22.
171. Lakdawala, S.S., Schwartz, R.A., Ferenchak, K., Carson, C.T., McSharry, B.P., Wilkinson, G.W. and Weitzman, M.D. (2008) Differential requirements of the C terminus of Nbs1 in suppressing adenovirus DNA replication and promoting concatemer formation. J. Virol. 82, 8362–8372.
172. Boyer, J., Rohleder, K. and Ketner, G. (1999) Adenovirus E4 34k and E4 11k inhibit double strand break repair and are physically associated with the cellular DNA-dependent protein kinase. Virology 263, 307–312.
173. Zhao, X., Madden-Fuentes, R.J., Lou, B.X., Pipas, J.M., Gerhardt, J., Rigell, C.J. and Fanning, E. (2008) Ataxia telangiectasia-mutated damage-signaling kinase- and proteasome-dependent destruction of Mre11–Rad50–Nbs1 subunits in Simian virus 40-infected primate cells. J. Virol. 82, 5316–5328.
174. Wu, X., Avni, D., Chiba, T., Yan, F., Zhao, Q., Lin, Y., Heng, H. and Livingston, D. (2004) SV40 T antigen interacts with Nbs1 to disrupt DNA replication control. Genes Dev. 18, 1305–1316.
175. Lilley, C.E., Carson, C.T., Muotri, A.R., Gage, F.H. and Weitzman, M.D. (2005) DNA repair proteins affect the lifecycle of herpes simplex virus 1. Proc. Natl. Acad. Sci. USA 102, 5844–5849.
176. Gregory, D.A. and Bachenheimer, S.L. (2008) Characterization of mre11 loss following HSV-1 infection. Virology 373, 124–136.
177. Dupre, A. et al. (2008) A forward chemical genetic screen reveals an inhibitor of the Mre11–Rad50–Nbs1 complex. Nat. Chem. Biol. 4, 119–125.

Standardized Cost Estimation for New Technologies (SCENT) - methodology and Tool

Stanil Y. Ereev and Martin K. Patel

Utrecht University, Copernicus Institute of Sustainable Development, Budapestlaan 6, NL-3584, CD Utrecht, Netherlands

ABSTRACT

This paper presents the development of a methodology and tool (called SCENT) to prepare preliminary economic estimates of the total production costs related to manufacturing in the process industries.

The methodology uses the factorial approach – cost objects are estimated using factors and percentages on the basis of the purchased equipment cost. The chosen approach is based on an extensive literature survey on methodologies and suitable data. The approach has the advantage that it can be based on a limited amount of data (list of equipment required for the technology). Therefore it is especially suitable for new or emerging technologies. The theoretical accuracy of the prepared estimates is within ±30%.

INTRODUCTION

Ever since the industrial revolution the process industries have played an important role in improving the quality of human life. Over the last 200 years, the process industries have gained significant importance in society by introducing products which dramatically changed the world. Key examples are pharmaceuticals, food products and food additives, fuels and polymers. At the same time, many technologies have led to controversial societal discussions, leading to the quest for a holistic technology assessment. There is a wide consensus that the three sustainability dimensions – economy, environment and society – need to be taken into consideration when assessing a technology.1 This paper deals with the (micro-) economic assessment, with the goal of preparing a readily applicable tool-set for the economic evaluation of new and emerging technologies.

In today's world, the economic performance is a conditio sine qua non for the existence and the future application of a technology. New technologies resulting in better, more environmentally friendly products are often more costly. The additional costs might be caused by higher capital investment (e.g. due to improved heat integration or measures for environmental or health protection) and/or higher operational costs. Decision making sometimes implies a trade-off between environmental or social benefits and economic costs. Therefore, careful and precise assessment of all costs related to a technology is of great importance to determine the prospects of a technology.

It is evident that there is a strong need for a reliable estimation of the full costs of a product on the micro level ("small" in Greek; refers here to the costs of a single manufacturing plant). These costs include many components, most of them not easy to forecast and calculate – capital investment in equipment and buildings, expenses for maintenance and repairs, materials and energy costs, salaries for employees.

Different approaches exist to obtain these data. For currently existing technologies, a relatively easy and trustworthy method is the comparison of historic real plant data. Many manufacturing processes have existed for more than 50 or 100 years and tens and hundreds of similar plants have been built on the planet. Such data are, however, not easily available due to their confidential and proprietary nature (e.g. Dysert, 2003). Moreover, when assessing new technologies, it is a further challenge, that historic data is typically of no or only very limited use.

Very often, estimates are made using complex commercial software tools.2 These tools may require a large amount of input data which are not always available for new technologies. Extensive previous knowledge and training on the software is also necessary. The commercial nature of the tools makes it difficult to compare estimates made by application of different tools. The need to rely on the outcome of one single tool is therefore quite common.

When preparing an economic analysis for new technologies, up-to-date data on prices are needed for many items, such as equipment, instrumentation and controls, chemicals, utilities (electricity, water, and natural gas), salaries for operating and skilled labour. One can obtain this information from vendors, suppliers, manufacturers, government statistics offices and others. All this requires significant effort. Unfortunately, there is no unified database which contains all necessary information for preparing basic cost estimates. Such a database was published in the open literature for the last time in 1990 (Couper, 2003).

There is hence an urgent need for a publicly available estimation method, which could provide sufficiently accurate results. It is apparent that in early stages of the development of new technologies, only study or preliminary estimates can be made. Despite the great deal of uncertainty existing for new technologies, the basic information required for conducting a cost analysis is often rather well known: this includes material balances (raw materials, solvents, catalysts), major pieces of equipment, important service facilities (e.g. steam generation) and energy balances (use of electricity, power).

Consequently, the question addressed by this study is how to use this information to arrive at reliable cost estimates for (new and emerging) technologies, thereby making use of existing cost estimation techniques. By conducting a literary survey covering different types of process cost analyses this paper pursues the following goals:

- To develop a cost estimation methodology to be used as standardized, default approach when making economic analysis for new or emerging technologies.
- To compile a database of all relevant costs and expenses necessary to prepare a study or preliminary estimate for a new technology; these include, for example, prices of major types of equipment, utilities, chemicals, environmental protection expenses and labour costs.
- To combine the methodology and the database into a simple cost assessment tool which would allow a quick and handy estimation and which can also be performed without previous specific expertise in cost estimation.

Many publications (mostly handbooks) deal with economic aspects of plant and process design, but most of them refer to just a handful of authors. From the literature review we performed we concluded that the most authoritative authors dealing with capital investment and production costs estimates are Peter, Timmerhaus and West (2004) as well as Couper (2003 and 2008). This paper is based mostly on their work but we have also accounted for the work of a few other authors (see section 3 "Essence of the factorial methodology").

CLASSIFICATION OF THE PRODUCTION COSTS

The classification in this study is mainly based on the work of Peters, Timmerhaus and West (2004). Their classification is the most comprehensive, includes most examples and was updated very recently. They are considered authorities in this area and many other publications refer to them. One important point made by them is that the main source of inaccuracy in economic estimates is actually not under-estimating or over-estimating individual data inputs rather than missing a cost object. Therefore the classification is of lower importance than the diligent application of the tool, thereby avoiding omissions. The classification presented in this paper (see below) exhibits an intermediate level of detail. In contrast, more elaborate classifications are applied when preparing detailed estimates. Examples of elaborate classifications are given by Couper (2003 and 2008) and by Holland and Wilkinson (1997).

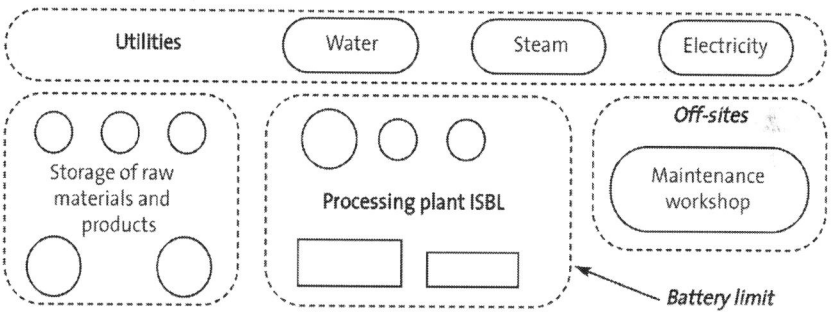

Figure 1: Example of a manufacturing plant: illustration of the processing plant (inside battery limits, ISBL) and the offsites (outside battery limits, OSBL). Adapted from Brennan (2004).

Production costs are the costs required for the plant to manufacture a product and they are expressed either per unit(s) produced (e.g. per tonne of product) or on basis of time: hourly, daily, monthly or annually. In effort of standardization, we recommend to express the production costs on an annual basis because it covers for seasonal variations in expenses, sales and process conditions as well as planned maintenance and shut-down periods. The production costs are generally classified as (semi-) variable and fixed costs. The variable costs are proportional to the load factor of the manufacturing, while fixed costs are independent of the plant capacity. Some of the variable costs are referred to as "semi-variable" because they have a minimum fixed component in them. The full list of production costs is given in Figure 2.

One of the most important fixed cost objects is the capital cost which includes all the buildings, machinery and equipment necessary for every-day operations of the plant. The initial capital investment (which may easily exceed $US 50 million for a large-scale manufacturing process; Peters, Timmerhaus and West, 2004) is recovered on annual basis through depreciation, with the depreciation regime being determined by tax laws. The annual cost is referred to as capital recovery cost, which is a fixed cost. The capital investment is generally classified in two main parts: the fixed-capital investment and the working capital.

The fixed-capital investment is all capital "fixed" to the ground, essentially tangible properties: e.g. equipment, machinery, buildings, and land. Since it may be as high as 80% of the whole capital investment (Peters, Timmerhaus and West, 2004) this cost might pre-determine the profitability of a technology and is a key factor in the decision making for a prospective investment.

For process industry plants, the fixed-capital investment can be divided in two parts: inside battery limits (ISBL or IBL) and outside battery limits or off-sites (OSBL or OBL). Battery limit is a real or imaginary geographical boundary around an area in the processing plant where the actual manufacturing takes place (the conversion of the raw materials or intermediates into the product). Thus the inside-battery limits costs may be defined as all expenses for equipment, including delivery, installation, foundations, structures, piping, electrical works, painting, insulation as well as the cost incurred for instrumentation, control equipment and operation. All these are direct costs. One could say that the battery-limits is a subsystem of the plant, with the raw materials (or intermediates) and the utilities flowing in and the products flowing out (Figure 1). Next to direct costs also indirect costs are applicable to the battery-limits, such as engineering and design expenses, construction costs, etc. These costs are typically charged to the project as a whole and cannot be assigned to specific cost objects. A list of the indirect costs is given in Figure 2. They normally include engineering and supervision expenses, construction costs for the project including the contractor's fee and all costs required to meet the legal requirements for building the plant. There is an additional "allowance" called contingency capital which is usually a percentage of the value of the whole project. This capital is meant to cover any unforeseen events, such as unpredicted delays due to weather conditions, strikes, transportation issues, etc.

The other direct costs (or the outside-battery limits costs) are expenses for land, yard improvements such as fences or roads, various buildings and service facilities (e.g. boilers, cooling towers, facilities for compressed air or steam generation). The latter are commonly referred to as "off-sites".

For every enterprise, there is also a sum of money required to conduct every day operations. It is necessary to cover expenses such as salaries, utility bills (electricity or natural gas), to regularly purchase raw materials and other supplies. This sum of money is called the working capital and it is not available for another purpose; therefore it is regarded as an investment item and is part of the capital investment.

Working capital is defined as the total amount of money invested in: raw materials and supplies in stock; finished products in stock and semi-finished products still in process of manufacturing; cash required for regular payments of operating expenses (salaries and other bills for a limited period); accounts receivable; accounts and taxes payable (Peters, Timmerhaus and West, 2004).

It is important to realize that even though operating expenses such as salaries, raw materials supply and others are taken into consideration in the working capital, the working capital is not an operating expense but that it is instead part of the capital investment. It is used to ensure liquidity of the firm. The reason behind this is that a company will have to constantly maintain cash to cover its every-day expenses.

The working capital is constantly regenerated with income from sales and stays at roughly the same level throughout the plant's lifetime. The working capital is by far smaller than the sum of all operating expenses for the whole year. It typically allows covering for one or two months of salaries, few months of raw materials supplies and other operating supplies. All this depends on the specifics of the business and the regularity of payments. It is also largely dependent on sales: seasonal sales will lead to less regular re-liquidation of the working capital and therefore, higher working capital will be required.

To facilitate the estimation of production costs, the SCENT tool (Standardized Cost Estimation for New Technologies) was developed. It has been prepared in the form of MS Excel file and is organized according to the classification presented above. It incorporates all equations and correlations between the cost objects which will be discussed in the further course of the paper. It follows a simple approach using drop-down menus and pre-defined values which allow usage without previous economic training.

(Semi-) Variable costs	Fixed costs			
Raw materials	Local taxes			
Operating labour	Insurance			
Direct supervisory and clerical labour	General plant overhead			
Utilities	Administrative costs			
Maintenance and repairs	Distribution and marketing			
Operating supplies	Research and Development			
Laboratory charges	Capital recovery - annualized percentage of Total capital investment (including interest)			
Patents, royalities				
	Total capital investment			
	Fixed-capital investment			Working capital
	Direct Costs		Indirect costs	
	Inside battery limits costs	Other direct costs	Engineering and Supervision	
	Equipment, incuding delivery	Buildings	Construction expenses	
	Equipment installation	Service facilities	Contractor's fee	
	Piping, electrical wors	Land	Legal	
	Insulation, painting	Yard works	Start-up capital	
	Instrumentation and controls		Contingency	

Figure 2: Classification of cost objects constituting production costs; Adapted from Peters, Timmerhaus and West (2004).

ESSENCE OF THE FACTORIAL METHODOLOGY

Based on a literature review it can be concluded that preliminary cost estimates are usually based on the cost of the purchased equipment, with all additional cost objects being estimated by means of specific default "factors", i.e. certain percentages of the purchased equipment cost. The accuracy of the estimate will vary depending on the level of detail known about the design of the plant. In early stages of the projects only preliminary estimates can be made. In later stages when there is more information about equipment requirements and the design specifications, more accurate estimates are possible (Towler and Sinnott, 2008). Important consideration is also the quality of the cost data (especially prices and scaling up or down for the specific technology).

From the preliminary flowsheets the pieces of equipment are selected and the purchased equipment cost is estimated. Usually prices of equipment are given by manufacturers and vendors as f.o.b.3. Delivery charges and installation expenses should then be added. The best source for this information are the manufacturers and sellers of the equipment, however, for preliminary estimates, such quotations might be too difficult or time-intensive to obtain (we have therefore compiled default data in the SCENT tool, see below).

From the cost of the delivered equipment the fixed-capital investment is determined. Once the fixed-capital investment has been estimated it is used as a base for estimating other costs: by multiplying the fixed-capital investment with different factors, one can obtain an estimate of the working capital and few major cost objects: e.g. maintenance, insurance, taxes and others (see below).

There are some costs which cannot be estimated on the basis of the purchased equipment cost or the fixed-capital investment since there is no correlation between them. Examples are the costs for raw materials and utilities. Those expenses are estimated from the material and energy balances (raw materials demand, utilities) and the respective prices or directly from the specifics of the technology (e.g. environmental expenses).

Towler and Sinnott (2008) give a short summary of the steps that need to be taken to make an estimate using the factorial methodology. The steps used in the SCENT tool follow a similar pattern (as mentioned above SCENT also contains databases on prices and estimation factors):

- From the preliminary flowsheets the estimator should identify the pieces of equipment required, together with specifics such as their capacity, material of construction (e.g. stainless steel versus regular steel), additional concerns such as extreme pressures or temperatures, etc.
- Determine the purchased costs and - by multiplication with equipment installation factors - estimate the installed cost for each piece of equipment (see below for further explanation)
- Estimate the fixed-capital investment including all direct and indirect costs 4.) Based on the fixed-capital investment estimate the working capital.
- The approach presented so far leads to the estimated total capital investment (see Figure 2). This cost occurs at the beginning of the project, depending on the time required for planning and construction. Afterwards, the actual manufacturing of the product starts and the total capital investment is gradually recovered. The capital recovery is included in the total production costs by application of the annuity method (see below). Production costs can be expressed as costs per year (or other time frame) or costs per unit of product (e.g. per tonne of product). For the reasons given above we determine in SCENT the costs per year.
- Based on the material and energy balances, estimate the raw materials (chemicals, solvents), utilities (water, electricity) and expenses for environmental measures
- Estimate the labour costs
- Estimate the total production costs including all (semi-)variable and fixed costs

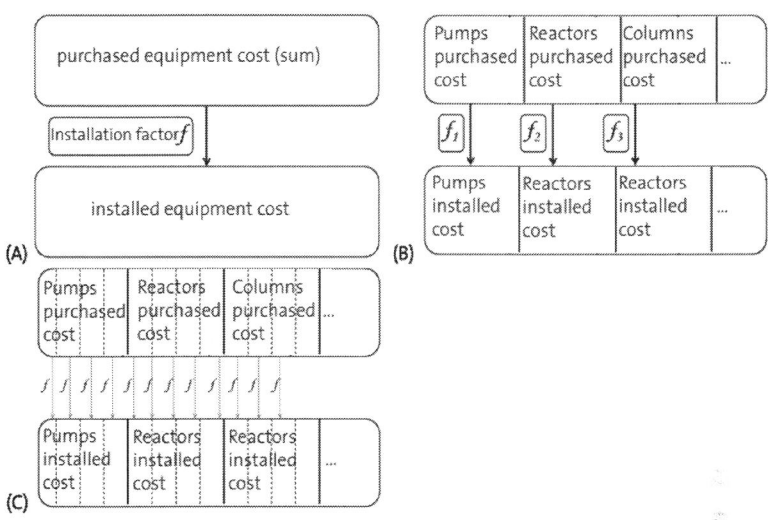

Figure 3: Illustration of the factorial approach: (A) single factor suggested originally by Lang (1948); (B) Group installation factors; (C) individual factors suggested by Woods (2008).

The most significant and also most difficult part in this process is the estimation of the fixed-capital investment. It relies mainly on the accuracy of: the input cost data of the purchased equipment, the installation factors used to account for the installation of the equipment (see below), and the correlation factors to account for all other direct and indirect costs (buildings, engineering, etc.). According to Peters, Timmerhaus and West (2004) the first two items may reach 80% of the fixed-capital investment. The equipment installation factors account for installation material and labour, foundations, structures, piping, fittings, electrical works, painting, insulation. Depending on the source of the factors, they might also include the required instrumentation and controls and might be based on the purchased equipment cost (f.o.b.) or on the delivered equipment cost (including freight charges). The value of the installation factor is always > 1, resulting in the installed equipment cost when multiplied with the purchased equipment costs. The accuracy of the input data can be increased by acquiring more recent, up to date information on the purchased cost of the equipment used. Many of the other cost objects are based on these cost data. Therefore increasing its accuracy will ultimately increase the quality of the final estimated cost. The best quality can be achieved by getting exact quotes from equipment manufacturers or suppliers – these are always recent, and will be chosen in accordance with the technical requirements of the process.

This factorial approach was first suggested by Lang in 1948. He differentiated between three types of processing plants: solid (e.g. a coal briquetting plant), solid-fluid (e.g. a shale oil plant with crushing, grinding, retorting and extraction) and fluid processing (e.g. a distillation separation system)4 and accordingly suggested three types of installation factors. He proposed the sum of all purchased equipment cost to be multiplied with the corresponding installation factor to yield as a result the sum of the installed equipment cost. This approach is illustrated in Figure 3(A).

Later on, these factors were refined by several authors (e.g. Hand, Wroth, and Guthrie (as quoted by Couper (2008)) and more specific "group installation factors"5 were developed as illustrated in Figure 3(B). More recently, individual factors have been developed. These installation factors are strictly specific for each individual type of equipment: i.e. two different types of pumps have different installation factors in contrary to the group factors where all types of pumps have the same installation factor. These factors are typically much more accurate than the previously presented methods. Woods (2008) gives a detailed list of about 500 different pieces of equipment, each along with individual installation factor. The individual factor approach is illustrated in Figure 3(C).

In the SCENT tool it was decided to use the individual equipment installation factors by Woods (2008) for the following reasons:
- Higher accuracy – the individual factors are more specific
- Most recent – they were first published in 2007 and re-printed in 2008
- Woods' approach deliberately excludes instrumentation and controls from its factors while earlier factors include it. The reason behind the exclusion is that instrumentation and

controls has undergone major development and accounting for them in a simplistic way by means of a default factor could therefore cause inaccuracies.
- Detailed – Woods also gives capacity exponents, alloy correction factors and additional correction factors for the temperature and pressure level and other process conditions which are all specific for each individual piece of equipment. This gives the estimator the opportunity for higher level of customization, and better accuracy of the estimate
- Labour / material ratio – Woods gives the ratio between the costs for labour and material which are incorporated in each installation factor. This makes it possible to correct in SCENT for the location by country, as will be described in detail in the next section

For all other cost objects in the capital investment and the production costs (e.g. off-sites and maintenance) mainly Peters, Timmerhaus and West (2004) are used as a source for few major reasons:
- Authoritative – the publication by this group of authors has been updated regularly and recently (2004). It has been referenced by many other authors working on the topic of cost estimation.
- Most consistent – many authors present factors to estimate certain cost objects, but the most comprehensive approach proved to be the one by Peters, Timmerhaus and West.
- Additional considerations – this group of authors presents additional considerations and different values for some of the cost objects (for example, they suggest three different values for buildings depending on whether a new plant is built on an undeveloped site or on an existing site or whether it is simply a small expansion on an existing site). These considerations allow for higher accuracy of the estimate. They will be presented in greater detail in the next section.

COST ESTIMATION

The process industries represent capital-intensive sectors (Economy Watch, 2010). Therefore, the accurate estimation of the capital investment is of crucial importance. When presenting the methodology the capital investment estimate will be discussed first, partially because it is the first step in project development but more importantly because the value of the capital investment is necessary to estimate other cost objects (e.g. the maintenance and repairs expenses).

Capital Investment Estimate

Purchased Equipment

In the SCENT tool the cost data by Woods (2008) have been implemented with all prices given f.o.b. in US $. The purchased equipment cost calculated by the tool includes in total all pieces of equipment from the process flow sheet, spare parts, surplus equipment, supplies and equipment allowance.

All prices in this database refer to a value of the Chemical Engineering Plant Cost Index (in short: CEPCI) of 1000. The CEPCI value for the years 1957–1959 was 100 while the value for 2010 was 585.9. By choosing today's (or an expected future) CEPCI value in SCENT, the results are adapted to the respective price levels. The CEPCI index is published at the end of each month in the Chemical Engineering magazine (Dysert, 2003).

The prices taken from Woods (2008) were published in the year 2007. In combination with the newest CEPCI values they allow to generate estimates with good accuracy also for a limited period in the future by using these historic cost data. Couper (2003) suggests that it is acceptable to use the same cost data with the correction of a cost index for no longer than 10 years. All prices in this database are given as base cost with a base capacity. The database also contains equipment-specific scaling exponents which allow estimating the cost of a given piece of equipment with a different capacity:

$$(\text{Cost equipment})_X = (\text{Cost equipment})_{base} \times \left(\frac{\text{CapacityX}}{\text{Capacitybase}}\right)^{EXP} \quad (1)$$

In case the scaling exponent is unknown, a value of 0.6 or 0.7 can be used as default (also referred to as six-tenths or seven-tenths rule). Equation (1) represents the economies of scale because buying a piece of equipment with twice the capacity is less than twice as expensive (when the exponent is less than 1.0). If, for a specific piece of equipment, this exponent is larger than 1.0, the most cost-effective way of scaling up is to duplicate the equipment. In the SCENT tool, a valid equipment-specific capacity range is suggested to the estimator.

Most of the equipment is offered with standard material of construction, usually cast steel (c/s) or cast iron. If special construction is required (e.g. stainless steel (s/s), nickel or any type of alloy), an alloy factor is applied to estimate the cost of the equipment. The alloy factor is specific for each type of material and equipment. Multiplication of this alloy factor by the cost of the equipment made of the standard material yields the cost of the equipment made from the chosen material as shown in equation (2).

Additional factors are provided to estimate the cost of equipment working at different process conditions or with different specifications: e.g. factors for elevated temperature or pressure. The approach is the same as with the alloy factors: the base cost of the equipment is multiplied with the additional factor (equation (2)). These factors are individual for each type and subtype of equipment (e.g. individual factors for pumps depending on the working pressure).

$$(\text{Cost equiment})_{PURCH} = f_{add} \times f_{alloy} \times (\text{Cost equipment})_{base} \quad (2)$$

Delivery charges for transportation, freight insurance, duties, and taxes are not accounted for by the installation factors. This cost may vary significantly depending on the plant's location or government regulations. Such estimation is hard to make in a very early stage of the project because the manufacturer or the vendor of the equipment might be yet uncertain as well as the plant's location might still be in question. Therefore, for preliminary estimates, 10% of the purchased equipment cost is proposed as an average, standardized value as delivery charge (Peters, Timmerhaus and West, 2004). The purchased equipment cost including charges for delivery will be referred to as "delivered equipment cost".

Installed Equipment

The installation factors account for installation material and labour, foundations, structures, process piping, pipe fittings, valves, painting, insulation, electrical systems and equipment switches, motors, feeders, grounding, wiring, lighting, panels, etc. The installation costs are estimated using equation (3):

$$(\text{Cost equiment})_{INST} = f_{INST} \times (\text{Cost equipment})_{PURCH} \qquad (3)$$

When the installed cost is estimated for equipment made of more expensive alloys, equation (3) needs additional correction because labour will generally be the same, structures and foundations will be mostly the same and also the expenses for electrical works might not be affected. When estimating equipment made of special materials, equation (3) therefore leads to overestimation. To correct for this, an alloy correction factor is applied according to equation (4):

$$(\text{Cost equiment})_{INST} = (1+(f_{INST}-1) \times f_{alloy\ corr}) \times f_{alloy} \times (\text{Cost equipment})_{base} \qquad (4)$$

The alloy correction factor is applied only when the installed cost is estimated and only when beforehand an alloy factor was used to estimate the purchased cost. The alloy correction factor is presented separately from the alloy factor, because it is only applied to the installed cost (this explains multiplication by the term

(f_{INST}-1) in equation 4), while the alloy factor is used to estimate the purchased cost of the equipment.

The alloy correction factor was introduced by Brown (2000) as modification to Hand's factorial approach (1958). The relation between the alloy correction factor and the alloy factor is given in Figure 4 6. It is logical that the higher the alloy factor, the higher the necessary correction is. In the figure the typical ranges for stainless steel and Monel alloy factors are presented, the exact values, however, remain specific for each type of equipment.

As mentioned above the installation factors presented by Woods provide the ratio between the labour and the material for each individual installation factor. This ratio could not be found in earlier publications. Using this ratio it is possible to develop simple country-specific factor to account for differences in labour costs in the various countries. The installation factor is split into two sub-factors, one for materials and one for labour costs as shown in equation (5):

$$f_{INST}-1 = Mf_{INST} + Lf_{INST} \qquad (5)$$

There are always differences between the installation and construction costs in different geographical locations (countries). Two important assumptions were made:

- Most of the large equipment, as used in the process industry, is globally traded. It is therefore justified to use uniform international prices for purchased equipment. In contrast, the labour costs related to installation and operation may differ significantly between countries and sometimes even within a country. To correct for this, SCENT assumes that the difference in costs between geographical locations (countries) is driven mainly by the difference in the labour rates (while the fluctuations in material prices were assumed to be negligible).
- The installation factors suggested by Woods are based mainly on US data; they take into consideration the US labour rates. When a piece of equipment is installed in the USA, the cost should be accurate; however, when the same piece of

equipment is installed outside the USA, the estimation might differ; it is assumed that this difference is driven solely by the difference in labour costs and therefore, a correction for this labour cost difference is applied.

Differences in labour costs are taken into account by introducing a correction factor which is applied only to the labor part of the installation factor. Subsequently, equations (3) and (5) are expanded into equation (6) to include this correction:

$$(\text{Cost equipment})_{INST} = (1 + (Mf_{INST} + Lf_{INST} \times f_{labour\ corr})) \times (\text{Cost equipment})_{PURCH} \quad (6)$$

Country-specific labour-related installation factors were created by normalizing the values for all European Union countries (plus Norway) relative to the factor for the USA which is 1.00 (Table 1). The country-specific factor indicates whether the installation of a piece of equipment is cheaper or more expensive compared to the USA, only from labour point of view. When this factor is multiplied with the labour component of the installation factor, it will decrease or increase the total cost.

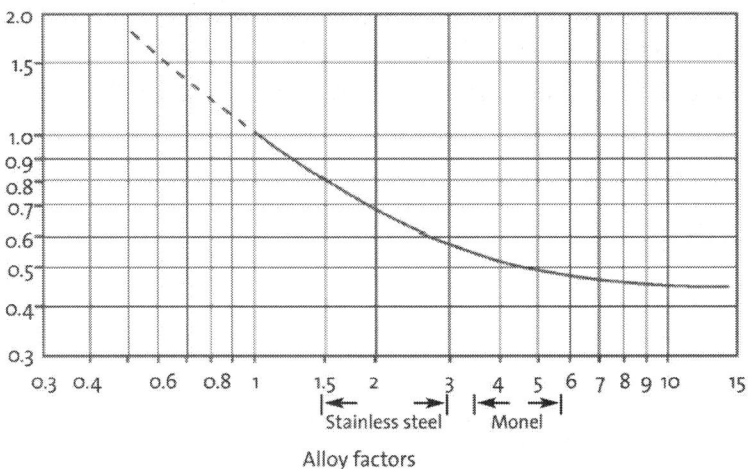

Figure 4: The alloy correction factors (on the y-axis) as a function of the alloy factor. Source: Woods (2008).

Fixed-capital Investment

The cost objects within the fixed-capital investment are commonly estimated by multiplication of the purchased or the delivered equipment with a suitable factor. It is important to point out that for many types of costs three different values will be suggested according to the three types of processing plants: solid, solid-fluid and fluid processing (see below Table 2; Peters, Timmerhaus and West, 2004). The various types of costs are discussed next.

The instrumentation and controls cost includes the purchased, delivered and the installed cost for any instrumentation and control equipment (such as alarms, sensors, valves, etc.), including all expenses for computer control and supportive software. In the SCENT tool there are two ways of estimating this cost, i.e. a detailed and a rough estimation method.

In the detailed estimation method, instrumentation and controls are considered as separate equipment and estimated based on a dataset including prices for control systems, sensors, alarms and others. The estimation is done in the same way as the estimation of the equipment cost. The source of the database is Woods (2008) and each piece of equipment comes together with individual capacity exponents, installation factors, alloy and additional factors. The purchased cost is linked to the CEPCI cost index and where applicable, exponents are used to correct for scale together with alloy and additional factors.

Table 1: Country-specific labour-related installation factors created by comparing labour rates. Calculated based on Eurostat and US Census data on labour costs in construction, 2009

Country	Labour-related Installation Factor	Country	Labour-related Installation Factor
Australia	1.03	Latvia	0.11
Belgium	1.04	Lithuania	0.16

Bulgaria	0.06	Luxembourg	0.99
Cyprus	0.61	Netherlands	1.28
Czech Republic	0.28	Norway	1.47
Denmark	1.10	Poland	0.19
Estonia	0.30	Portugal	0.34
Finland	1.09	Romania	0.09
France	1.05	Slovakia	0.20
Germany	0.93	Slovenia	0.51
Greece	0.46	Spain	0.73
Hungary	0.19	Sweden	1.14
Ireland	1.35	United Kingdom	1.06
Italy	0.76	USA	1.00

In the rough estimation method, the factors suggested by Peters, Timmerhaus and West (2004) are used. The corresponding factor, depending on the type of processing plant, is multiplied with the delivered equipment cost, to represent the capital costs related to the instrumentation and controls (Table 2).

The cost object of buildings represents all expenses for process-related buildings (including sub- and superstructures, platforms, stairways), auxiliary buildings (e.g. administration and office, garage, product or spare parts warehouse, safety and security, fire station, shipping office, research and control laboratories), maintenance shops (e.g. electric, piping or machine). This cost includes all necessary building services – plumbing, heating, ventilation, lighting, elevators, telephones, intercommunication systems, painting, fire alarms and others.

The cost of buildings is estimated by multiplying the purchased equipment cost (excluding delivery) with a representative factor. Different values are presented by Peters, Timmerhaus and West (2004) based on three additional considerations: whether a new plant is built at a new site, a new unit is built at an existing site or simply an expansion is made to an existing plant (Table 2). The cost of buildings is lowest in the case of expansion, because most of the required infrastructure already exists and the cost is highest in the case of building new plant at undeveloped site.

The service facilities are all additional process or non-process equipment which are not directly involved in the manufacturing of the end product but are of crucial importance for the whole plant operations. They are commonly referred to as "off-sites" and fall under the category "outside-battery limits". Brennan (2004) points out that mostly the term "services" is used as synonym for utilities and the term "plant service facilities" also includes buildings.

These include facilities required for the supply of utilities such as steam, water, power, refrigeration, compressed air, and also waste disposal facilities. Other types of facilities which could fall under this cost object are, for example, water treatment and storage, cooling towers, electric substation, air separation plant, fuel storage, waste disposal plant, environmental controls and fire protection. Non-process equipment required for the plant is also estimated as part of this item – office furniture and equipment, shelves, bins, safety and medical equipment, fire extinguishers, hoses and engines, loading stations, and important distribution and packaging equipment – raw material and product storage and handling equipment, blending facilities, etc.

There are three ways of estimating this cost: detailed and rough factorial estimation and through selecting pieces of equipment.

- In the detailed factorial estimation, the estimator is presented with a short list of some of the more common service facilities at manufacturing sites (Appendix Table A1). One can choose between low, typical or high values of percentages of the fixed-capital investment for each of the given facilities.
- The rough factorial estimation method is based on the delivered equipment cost using factors suggested by Peters, Timmerhaus and West (2004) in Table 2.
- As alternative to the factorial method it is also possible to extract values for the major pieces of equipment required in the service facilities from the cost database (this is possible because Woods (2008) gives cost data for such equipment, e.g. steam turbine or waste water treatment unit).

Engineering and supervision costs include expenses for administration, process design and general engineering, computer graphics, cost engineering, communications; also consultant fees, travel expenses, as well as engineering supervision and inspection.

The category for construction expenses and contractor's fee comprises all costs for construction, operation, also maintenance of temporary facilities, offices, roads, communications and fencing; all expenses for construction tools and equipment, supervision, accounting, timekeeping and purchasing. Additional costs such as warehouse personnel and expenses, guards, safety, permits, taxes, insurances and interest are estimated as part of this cost object. Peters, Timmerhaus and West suggest separate factors for estimating the contractor's fee.

In order for the project to be executed, some legal costs are incurred, such as the expenses for the process of identification of applicable federal, state and local regulations, preparation and submission of forms required by regulatory agencies, the process of acquisition of regulatory approval and contract negotiation costs.

When executing new projects, there is high possibility of certain unforeseen events to occur that might cause delays or additional costs. Possible examples are extreme weather conditions (storms, floods), transportation accidents, strikes by transportation or construction personnel, design changes, omissions, errors or inaccuracies in estimation as well as various construction problems. The costs related to such unforeseen events are covered by the contingency capital, which is sometimes also referred to as back-up capital.

It is important to note that compared to other sources (e.g. Couper or Woods) the contingency capital values suggested in Table 2 are on the low side as other sources suggest values around 15–25% of the fixed-capital investment. Therefore, these values are meant for orientation, bearing in mind that for the respective technology studied, different values might be more appropriate. For new technologies the estimator might prefer much higher values, depending on the technology features. It is generally accepted as a rule of thumb that the lower the total value of the project, the higher the contingency capital must be (Table 2).

The land and yard improvements costs include all necessary capital for land surveys and fees, the property cost, and yard improvements such as expenses for site development (site clearing, grading) and landscaping, roads, walkways, railroads, fences, parking areas, etc.

Land and yard improvements are sometimes excluded from the capital investment. Land is considered to be completely recoverable at the end of the plant's life and therefore does not need to be capitalized. The value for the land cost is suggested to be between 4% and 8% of the delivered equipment cost, with 6% being an adequate average value (Table 2).

The yard improvements are considered to increase the value of the land, so again they do not need to be capitalized (as recoverable) (Silla, 2003). They are estimated using two alternative sets of factors (see Table 2); the first is based on Peters, Timmerhaus and West (2004) and refers to the delivered equipment cost; the second source originates from Silla (2003) and is based on the fixed-capital investment. The first approach (based on the delivered equipment cost) tends to give results that are on the higher side.

Table 2: Factors for estimation of the direct and the indirect costs of the fixed-capital investment; Source: Peters, Timmerhaus and West (2004)

Type of cost	solid	solid-fluid	fluid
Instrumentation and Controls	0.18	0.26	0.36
Buildings[†] - new plant at new sites	0.68	0.47	0.45
Buildings - new unit at existing site	0.25	0.29	0.11
Buildings - expansion at existing site	0.15	0.07	0.06
Service facilities	0.40	0.55	0.70
Engineering and Supervision	0.33	0.32	0.33
Construction Expenses	0.39	0.34	0.41
Contractor's fee	0.17	0.19	0.22
Legal Expenses	0.04	0.04	0.04
Contingency	0.35	0.37	0.44
Land	0.06	0.06	0.06
Yard improvements - higher	0.15	0.12	0.10
Yard improvements[‡] - lower	0.0285	0.0249	0.0211

*All factors are based on the delivered equipment cost except in the cases of:

†factors for buildings are based on the purchased equipment cost;

‡factors for the lower estimate of the yard improvements are based on the depreciable fixed-capital investment; Source: Silla (2003).

Before the plant reaches its regular operational mode it runs through a so-called "start-up" period. During this time, additional expenses will occur such as allowance for testing and adjustment of equipment, piping and others, the calibration of instrumentation and control equipment. The start-up costs include materials, utilities and labour for checking and testing the plant and initial personnel training. These costs are considered part of the capital investment.

There are two approaches for the estimation of this cost: the single- and the multiple-factor approaches both suggested by Couper (2003). The single-factor approach is a percentage of the fixed-capital investment credited to the start-up expenses. The multiple-factor is based on a few items like the number of months for training of the personnel and for the start-up of production, initial inefficiency expenses as percentage of the production costs and others. The multiple-factor approach will not be used in SCENT as it requires data which might not be available for new technologies.

The single-factor approach for estimating the start-up expenses gives three values of percentages (6, 8% and 10%) of the fixed-capital investment depending on the magnitude of the investment. The factor given in Appendix Table A2 is multiplied with the value of the fixed-capital investment to determine the start-up estimate.

Working Capital

By analogy with the estimation of the start-up capital, there are two widely accepted approaches for the estimation of the working capital: the percentage and the inventory method (Couper, 2003). The inventory method takes into consideration few major items: raw materials cost and periods of supply, semi-finished and finished products cost, period of sales and others. Such detailed information is likely to be unavailable for a new technology; therefore, the chosen percentage method is presented next.

The percentage method simply estimates the working capital as a percentage of another cost and can be based on either the capital investment or the annual sales. Since for new technologies, there might be a great uncertainty with regard to the sales, the percentage of capital investment was selected as more suitable option. The working capital is between 15% and 30% of the total capital investment. The total capital investment includes two components: the fixed-capital investment and the working capital (see Figure 2). Once the fixed-capital investment is estimated, the working capital is estimated on its basis.

As explained above the working capital is constantly regenerated by income from sales; therefore it strongly depends on the type of sales rate. If a product is sold at relatively constant and uniform yearly rate, then the regeneration of working capital has smaller fluctuations and values between 15% and 25% of the total capital investment are suggested for the working capital. If the product has very high seasonal variations in sales, then higher values are proposed: 20%–30% of the capital investment (Couper, 2003).

Production Cost Estimate

As explained above, in SCENT the final production cost estimate is on annual basis. The capital investment cost is annualized to represent the yearly capital recovery expense. The amounts of raw materials required for the manufacturing of the product are estimated based on the material balance. The category "raw materials" consists mostly of chemicals, catalysts and solvents. A database of prices for more than 700 types of chemicals is incorporated in the SCENT tool. The source of these prices is the www.icis.com website (free-of-charge area). These prices were initially published in the 28 August 2006 issue of the Chemical Market Reporter magazine (now existing as ICIS Chemical Business). Most of the prices are from 2006, while some of these prices were recently updated in 2007 and 2008. The prices in this database are given in US $ and in US customary units (not SI units) – e.g. gal, lb. For this reason – a unit convertor is included in the SCENT tool. For most of the prices, the geographic origin of the price is given and it is very likely that in different areas of the world, the price would be different. For this reason the cost data should be used with caution, the prices in the embedded database serves for orientation and preliminary estimates but wherever possible, more accurate data e.g. from quotations made by possible vendors and suppliers should be used.

The expenses for utilities such as diesel oil, gasoline, natural gas, electricity and water are also estimated from the material and energy balances. A database of prices was compiled mainly from Eurostat and the US Energy Information Administration, with all of the prices being country-specific (the prices are included in the SCENT tool).

The cost data for electricity and natural gas are valid for industrial consumers. Country-specific water prices are difficult to obtain. The source used for water prices in SCENT is a 2008 report by NUS Consulting. All of the water prices are unfortunately for household consumers, and some of the countries have country-specific prices, while for the rest – the European average is used. The estimator should note that the prices for household consumers are typically much higher than the local prices for industrial consumers. For preliminary economic estimates, such rough values of water prices are considered acceptable, but they should be replaced by more accurate data when preparing a more detailed estimate.

The labour costs are estimated in two parts: operating labour costs and direct supervisory and clerical labour costs. The direct supervisory and clerical labour costs are estimated to be between 10 and 20% of the operating labour costs, with 15% being an average value (Table 3).

The operating labour is estimated by multiplying the number of required employees by the average labour cost in the manufacturing industry in the different countries. The number of required employees could be estimated in three alternative ways:

- Based on the type of equipment. Ulrich (1984) developed a table of the most common types of equipment and assigned a representative number of operators per shift to each type of equipment. For example a heat exchanger requires 0.1 operators per shift and a cooling tower requires 1 operator per shift (the specific values are given in the SCENT tool). This allows estimating the total number of operators required per shift for the plant operations. It is assumed that an employee works 5 shifts of 8 hours per week, for 48 weeks per year. Then, on the base of the load factor of the technology, the required number of full-time employees is estimated.
- Peters, Timmerhaus and West (2004) suggest representative values of required employee-hours for manufacturing 1000 kg of end product. For solid-processing plant, the values are between 4 and 8, for solid-fluid processing plant: between 2 and 4 and for fluid-processing plant the suggested values are between 0.33 and 2. On the base of the capacity and the load factor of the technology, the necessary number of employees is estimated.
- Wessel (1952) developed an equation to estimate the labour requirements for production rate of 2000 (short) tons/day (1814 metric tonnes/day). The method gives the number of operator-hours per ton per processing step. A processing step is defined as a step in which a unit operation occurs (Couper, 2003), e.g. filtration or distillation are considered separate steps. Equation (7) below is adapted from Couper (2003):

$$\log Y = -0.783 \log X + 1.252 + B \qquad (7)$$

Y is the operating labour in operator-hours per ton (short ton) per processing step

X is the plant capacity in tons (short tons) per day

B is a constant depending on the type of process: + 0.132 (for batch operations), + 0 (for operations with average labour requirements), − 0.167 (for well-instrumented continuous process).

Once the number of the required employees is estimated through one of the three approaches, this number is multiplied with the country-specific labor cost. The specific salary rates are given in the SCENT tool with the respective sources, namely Eurostat and the US Census Bureau.

The approaches 2 and 3 presented here are based on historical data from the chemical industry and are commonly used as a rule of thumb for preliminary estimates. Ulrich's approach might be more accurate for new technologies because it is not based on historical data, but on equipment specifics. For this reason it is recommended for new technologies and the other two approaches are simply given as alternatives. As a further argument put forward by Couper (2008) is that Ulrich's approach is also simpler.

According to Peters, Timmerhaus and West (2004) the maintenance and repairs costs vary depending on the type of chemical process: for a simple chemical process, the maintenance costs are low (2-6%) and tend to rise for an average process (with normal operating conditions: 5-9%) and for a complicated process (or with severe corrosion operating conditions, or with extensive instrumentation: 7-11%). The maintenance and repairs costs refer to the fixed-capital investment and have two parts, namely labour and material. The corresponding factors are given in Appendix Table A3 for a low, average and high cost level for each component.

The costs for operating supplies include expenses for lubricants, test chemicals and spare parts and they are estimated at 10 to 20% of the maintenance and repairs costs (Table 3).

Other (semi-) variable costs include the laboratory charges and expenses for patents and royalties. The laboratory charges are estimated at 10 to 20% of the operating labour (Table 3). Another important cost object are the patents and royalties (0 to 6% of the total product cost is suggested by Peters, Timmerhaus and West, 2004). For new or emerging technologies, however, typical percentages might not be correct and that is why this cost might be very specific to the technology in question. For this reason the costs for patents and royalties were excluded from SCENT.

The factors used to estimate the fixed production costs are presented in Table 3 together with the quantity they refer to (second column from the left). Local taxes are likely to differ depending on the location of the plant – and they are estimated at 1-2% of the fixed-capital investment in less populated areas and at 3-4% in more populated areas. Insurance is accepted to be roughly 1% of the fixed-capital investment (or less).

The general plant overhead comprises all costs for general plant upkeep, packaging, medical services, safety and protection, storage facilities and others. Important part of this cost is the payroll overhead which is suggested to be typically between 30 and 40% of the labour costs (Couper, 2008). Administrative costs include executive salaries, clerical wages, legal fees, office supplies and communication. These costs are estimated around 15-25% of the operating labour. Distribution and marketing expenses are typically spent on sales offices and salespeople, shipping and advertising (Table 3).

Research and Development costs are excluded from SCENT as for new and/or emerging technologies these might be atypically high and a preliminary estimate might be inaccurate.

Table 3: Factors for estimation of some (semi-) variable and fixed production costs

Type of cost	Based on:	low value	average value	high value
Direct supervisory & clerical labour	Operating labour	1.10	0.15	0.20
Operating supplies	Maintenance	0.10	0.15	0.20
Laboratory charges	Operating labour	0.10	0.15	0.20
Local taxes - less populated area	Fixed-capital investment	0.01		0.02
Local taxes - more populated area	Fixed-capital investment	0.03		0.04
Insurance	Fixed-capital investment	0.01	0.01	0.01

General plant overhead	Labour & Maintenancet	0.50	0.60	0.70
Administrative costs	Operating labour	0.15	0.20	0.25
Distribution and marketing	Total procution cost	0.02	0.11	0.20

The initial capital investment is recovered through depreciation which depends on the discount rate and the lifetime of the manufacturing plant. The applicable depreciation regime is likely to differ by country. Financing (in terms of interest on borrowed capital) must also be included as an annual cost. For new technologies, financing, interest rates, government subsidies, etc. are subject to high uncertainties. For this reason, a strongly simplified method will be used in SCENT. A capital recovery factor α is calculated according to equation (8):

$$\alpha = \frac{r}{1-(1+r)^{-L}} \qquad (8)$$

where: α – capital recovery factor (annuity factor), r – interest rate and L – capital recovery period (in years). The annual capital recovery is calculated according to equation (9):

(Capital recovery)=α×(Total capital investment) (9)

ACCURACY OF THE METHODOLOGY

All factors presented above are based on actual manufacturing plants. It is reported that estimates obtained with them provide uncertainty of the results at ±30% (Couper, 2008). Woods as well as Peters, Timmerhaus and West also suggest the same theoretical accuracy for the cost data provided for the purchased equipment

114   Machinery Component Maintenance and Repair

and the factors proposed for estimating the remaining costs. Therefore, it is assumed that the theoretical accuracy of the SCENT tool is also expected to be around ±30%. In order to gain first insight into the accuracy of SCENT, it was validated for three types of manufacturing processes. Due to the confidentiality of the data used, the exact names of the products and overall costs are not presented and only the inaccuracy of this methodology against the cost value reported in the original source is shown.

It was found that the biggest source of inaccuracy is the off-site capital and this is acknowledged as a limitation of this methodology. Therefore the validation was adjusted to exclude the estimation of off-sites and the respective deviations are also given in Table 4.

The main factors affecting the final accuracy are the quality of the input cost data and the accuracy of the factors to estimate the different cost objects. A cost assessment of higher quality can be achieved by investing more resources in obtaining more accurate input prices of equipment as many other cost objects are based on the cost of the purchased equipment.

It is important to note that for preliminary purposes, 30% accuracy is quite sufficient and widely accepted in literature (see Couper, 2003 and 2008). A cost assessment of this type is not meant to serve as basis for a final conclusion on the economic viability of a technology. Instead, the ultimate purpose of such preliminary estimate is to formulate a recommendation whether the technology is promising and whether therefore a more accurate estimate using better input data should be prepared, e.g. by application of commercial software tools (see footnote 2) or in close cooperation with equipment suppliers and engineering companies offering turn-key plants.

Table 4: Results from validation of the SCENT tool in % as compared to reference values (confidential)

Manufacturing of:	Fixed-capital investment incl. off-sites	Fixed-capital investment excl. off-sites
Alcohol	-19%	-10%

| Organic acid i | -29% | +1% |
| Organic acid 2 | -22% | -7% |

CONCLUSIONS AND DISCUSSION

The methodology presented in this paper and implemented in the SCENT tool offers a comprehensive approach for preparing preliminary economic estimates for plants operated by the process industries. The SCENT tool was developed in the form of a MS Excel file which is simple to use and is publicly available at www.prosuite.org. SCENT focuses particularly on new or emerging technologies for which the available data is typically scarce and/or uncertain.

The methodology uses the factorial approach – cost objects are estimated using factors and percentages on the basis of the purchased equipment cost. The chosen approach is based on an extensive literature survey on methodologies and suitable data. SCENT is operated on a limited amount of data (list of equipment required for the technology). Therefore it is especially practical for new or emerging technologies.

The most important cost item in the estimation process is the purchased and the installed equipment cost, mainly because it is a major part of the fixed-capital investment (can reach up to 80% of it) and also because it is used as a base for the estimation of the remaining cost objects. Against this background, the individual installation, alloy, capacity and additional factors presented by Woods (2008) were used in SCENT since they increase the accuracy of the estimate.

A limitation of the methodology is the fact that all factorial correlations originate from plants and processes based in the USA. Since the process industry is globalized and typical equipment, materials, and design specifications, etc. are similar throughout the world, the selected approach is likely to produce sufficiently accurate preliminary results also for plants operated elsewhere. While material costs can also be assumed to be globally comparable, there are substantial differences in labour rates, even between countries from the same region. Therefore, a labour-related installation correction factor is introduced in this paper which accounts for the differences in labour rates among the European Union countries, as well as Norway and the USA. It increases the accuracy of the capital investment estimate.

A database has been compiled with recent prices and costs for nearly 500 pieces of equipment, selected utilities, over 700 types of chemicals and the labour costs in most European countries.

This database also includes a short list of typical environmental protection expenses. It is acknowledged in this paper that this list is not representative and insufficient to estimate all environmental protection expenses which might occur for a technology.

More data would be required in order to reflect more accurately the differences between the countries: country-specific costs for environmental protection expenses or taxes, infrastructure and transportation-related costs (currently delivery charges are assumed to be fixed at 10% of the purchased equipment cost, despite location), regulations, local laws, labour productivity and others.

In effort to develop a standardized approach, few cost items were deliberately excluded from the analysis because they might vary substantially for new or emerging technologies: costs for research and development, patents and royalties, subsidies or interest rates. Possible technological learning with time is also neglected. Technological learning can, however, be incorporated in SCENT when estimating the capital investment. If projects which are similar to the emerging technology have already been executed, one can review the different cost objects and conclude whether some of them showed substantially higher or lower final costs as compared to initially estimated. The methodology also allows for accounting for expected higher yields and the improvement of any other parameter, but this is again technology-specific and cannot be included in a standardized approach.

This methodology is hence capable of offering reliable estimates for preliminary purposes. This was confirmed by applying the SCENT tool for three types of manufacturing processes. The resulting estimates of the fixed-capital investment were within the expected accuracy range, with the highest inaccuracy observed for the off-site capital. This is acknowledged as a limitation of this methodology. Based on the validation runs, the literature information on the method and the quality of the data used, we conclude that the results generated with the SCENT tool have a theoretical uncertainty of ±30%.

In conclusion, SCENT is recommended as suitable approach to estimate the production costs of new or emerging technologies.

Appendix

Table A1: Factors for estimation of the labour and material parts of the maintenance and repairs costs. Adapted from Peters, Timmerhaus and West (2004)

	Labour			Materials		
Type of process:	low value	average value	high value	low value	average value	high value
simple chemical process	0.01	0.02	0.03	0.01	0.02	0.03
average process	0.02	0.03	0.04	0.03	0.03	0.05
complicated process	0.03	0.04	0.05	0.04	0.04	0 06

Table A2: Percentages of the fixed-capital investment for major service facilities. Adapted from Peters, Timmerhaus and West (2004)

type of facility	low value	typical value	high value
steam generation	2.6	3.0	6.0
steam distribution	0.2	1.0	2.0
water supply, cooling and pumping	0.4	1.8	3.7
water treatment	0.5	1.3	2.1
water distribution	0.1	0.8	2.0
electrical substation	0.9	1.3	2.6
electrical distribution	0.4	1.0	2.1
gas supply and distribution	0.2	0.3	0.4
air compression and distribution	0.2	1.0	3.0
refrigeration including distribution	0.5	1.0	2.0
process waste disposal	0.6	1.5	2.4
sanitary waste disposal	0.2	0.4	0.6
communications	0.1	0.2	0.3
raw material storage	0.3	0.5	3.2
finished product storage	0.7	1.5	2.4

| fire protection system | 0.3 | 0.5 | 1.0 |
| safety installations | 0.2 | 0.4 | 0.6 |

Table A3: Factors for estimating the start-up capital. Factors are based on the fixed-capital investment, Adapted from Couper (2003)

fixed-capital investment	factor
> mo milliopn US $	0.06
10 —loo million US $	0.08
< up million US $	0.10

REFERENCES

1. Blok, K. (2008): *Introduction to Energy Analysis*, Techne Press, Washington D.C.
2. Brennan, D.J. (2004): *Process Industry Economics: An International Perspective*, IChemE, London.
3. Couper, J.R. (2003): *Process Engineering Economics.*, Marcel Dekker, New York.
4. Couper, J.R. (2008): Process Economics section, in: Green, D.W., Perry, R.H., (Eds.) Perry's *Chemical Engineering Handbook*, 8th ed., McGraw-Hill, Columbus (OH).
5. Dysert, L.R. (2003):. Sharpen Your Cost Estimating Skills, *Cost Engineering*, 45 (6), 22-30.
6. Economy Watch (2010): retrieved 6-Sep-2010, http://www.economywatch.com/world-industries/capital-intensive.html.
7. Guthrie, K.M. (1969), *Chemical Engineering*, 114–156.
8. Guthrie, K.M. (1974): Process Plant Estimating, Evaluation and Control, Craftman Book Company of America, Solana Beach, CA.

9. Hand, W.E. (1958): *Petroleum Refiner*. September 1958:331–334.
10. Holland, F.A., Wilkinson, J.K. (1997): Process Economics section, *Perry's Chemical Engineering Handbook*, 7th ed., McGraw-Hill, Columbus (OH):.
11. Peters, M.S., Timmerhaus, K., West, R. (2002): *Plant Design and Economics for Chemical Engineers*, McGraw-Hill, Columbus (OH).
12. Silla, H. (2003): *Chemical Process Engineering* (Design and Economics), Marcel Dekker, New York
13. Towler, G., Sinnott, R.K. (2007): *Chemical Engineering Design: Principles, Practice and Economics of Plant and Process Design*, Butterworth-Heinemann, Oxford.
14. Ulrich, G.D. (1984): *A Guide to Chemical Engineering Process Design and Economics*, John Wiley & Sons Inc, Hoboken, NJ.
15. Wells, G.L., Rose, L.M. (1986) *The art of chemical process design*, Elsevier Science Pub, Amsterdam.
16. Woods, D.R.(2008): *Rules of Thumb in Engineering Practice*, Wiley-VCH, Hoboken, NJ.
17. Wroth, W.F. (1960.): *Chemical Engineering*, 1960:204.

Chapter 4

Removal of Greases and Lubricating Oils from Metal Parts of Machinery Processes by Subcritical Water Treatment

Walter J. Weber Jr[a] and Han S. Kim[b]

[a]Environmental and Ecological Sciences and Engineering, Department of Chemical Engineering, University of Michigan, 4103 Engineering Research Building I, Ann Arbor, MI 48109-2099, USA

[b]Department of Environmental Engineering, Konkuk University, 1 Hwayang-dong, Gwangjin-gu, Seoul 143-701, Republic of Korea

ABSTRACT

Toxic or persistent solvents have been widely used to remove greases and lubricants from various machine elements in the washing processes. In this study, an alternative degreasing method that employed subcritical state water was assessed. This environmentally benign solvent has significant potential for various degreasing applications. The operation time and temperature and flow rate of subcritical water had markedly positive impacts on the degreasing efficiency. However, the effect of pressure of subcritical water flow was minimal. The degreasing efficiency was also highly dependent on the physical characteristics and chemical composition of grease. The subcritical water treatment demonstrated a better degreasing efficiency than conventional degreasing methods. Only minor physical damage was observed on the metal parts after the subcritical water treatment. Conclusively, it was found that the subcritical water degreasing system can be used as an effective degreasing technology for machinery operations.

GRAPHICAL ABSTRACT

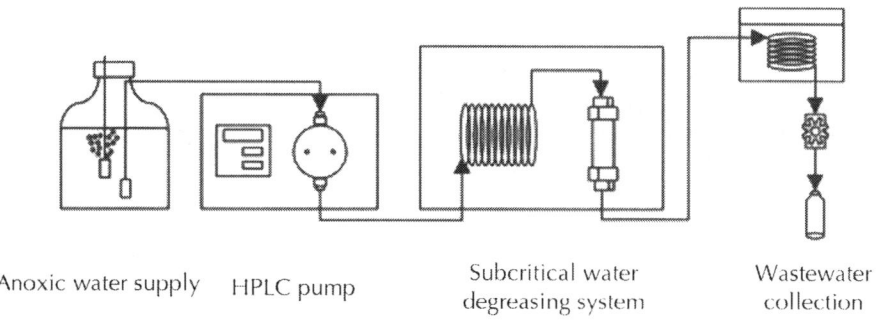

Anoxic water supply   HPLC pump   Subcritical water degreasing system   Wastewater collection

INTRODUCTION

Greases and lubricating oils are ubiquitously used in machine-related processes and operations, providing a fluid layer to separate moving surfaces of machines, minimizing heat and friction, and hindering surface wear under extreme temperature and pressure conditions [1]. These materials are typically mixtures of alkanes, alkenes, alicyclics, and aromatics distilled from crude oil and then modified with various thickeners and other additives to impart the characteristics required for specific applications [2] and [3]. Proper removal of these products from the metal parts after use is necessary for the continuous and repetitive operation of machinery processes [4]. It is, in fact, not easy to remove greases and lubricating oils from metal surfaces during production or machinery repair since they are complex mixtures designed to resist degradation during strenuous use [5].

Typical multipurpose greases contain about 85% base oil, 10% thickener, and 5% additives [1]. The most common base oils are mineral oils, but synthetic oils are also popular for special grease applications. The chemical composition of mineral oils is not easily defined because they can contain various hydrocarbons depending on the sources of crude oils specific to origin and extent of refining. Synthetic hydrocarbons such as polyalphaolefins, alkylated aromatics, polyglycols, ester and silicone oils, and perfluoropolyethers, are more clearly defined with regard to their molecular weight and structure. The second most abundant component of grease is the thickener. Thickeners are finely dispersed in the base oils, forming a structure similar to an emulsion and these substances primarily determine the physical property of the grease [2]. Soaps of lithium, calcium, sodium, or aluminum are the most common thickeners, but solid particles such as bentonite clay, silica, graphite, and polymers are also employed. The third main components of grease are additives, including anti-oxidants, anti-wear additives, and friction and structure modifiers.

Conventionally, combinations of cleaning and rinsing baths, mechanical treatment (ultrasound, spraying, and injection flood washing), and drying (hot air, vacuum, and denaturing fluids) have been used, depending on specific applications and the nature of the items that require decoating [3] and [5]. The most common solvent media used for degreasing can be classified as follows: organic solvents, strong alkalies (e.g., sodium hydroxide), alkali salts (e.g., cyanides and borates), complexing agents (e.g., EDTA and NTA), surfactants (e.g., nonylphenyl), and corrosion inhibitors (e.g., sodium benzoate). Among them, organic solvents are known to be hazardous to both human health and the environment, and in addition, some of them are potential or known carcinogens [5]: e.g., hexane, trichloroethane, trichloroethylene (TCE), tetrachloroethylene, chloro-fluoro-carbon 113, and dichloromethane (DCM). The aqueous-based solvents are usually strongly alkaline or acidic in nature and therefore highly toxic and corrosive [3].

Subcritical water represented by superheated liquid (water heated to any temperature less than its critical temperature of 373 °C with enough pressure to maintain its liquid state) has been explored in a number of prior studies [6] and [7]. This environmentally benign solvent has significant potential for beneficial use in a wide variety of degreasing and decoating applications [8]. Subcritical water can dissolve polar and non-polar organic compounds that are insoluble under ambient conditions because its dielectric constant (ε) decreases from ~80 at room temperature to ~30 at 250 °C [9] and [10]. At 300 °C, the density and ε of water approach those of ambient acetone [10]. As a result, the solubilities of hydrophobic organic compounds in the subcritical water can be increased significantly by raising the temperature [11] and [12]. Another benefit of subcritical water for degreasing is the potential for recovery, recycling, and reprocessing of valuable organic chemicals, organic solvents, and greases [13].

In this study, the development of an alternative degreasing method that employs essentially pure water was explored. Specifically, the removal of grease and lubricating oil products from metal parts using a subcritical water extraction system was evaluated. Tests were performed on a variety of greases and lubricating oils under varying operating conditions. The degreasing efficiency of the subcritical water treatment was compared with those obtained by using conventional degreasing and cleaning agents. We also investigated the corrosion potential of subcritical water for future use. The results of this study are expected to demonstrate the general applicability of the subcritical water degreasing system.

MATERIALS AND METHODS

Materials

Greases and lubricating oils, chosen based on their application and components, were purchased from local mechanic shops (Table 1). Organic solvents (hexane, DCM, TCE, and methanol) and Zep cleaner used as a conventional degreasing agent were purchased from Fisher Scientific (Chicago, IL, USA) and Zep Superior Solutions (Cartersville, GA, USA), respectively. Petroleum hydrocarbon standard and alkane hydrocarbons (pentadecane [$C_{15}H_{32}$], eicosane [$C_{20}H_{42}$], and triacontane [$C_{30}H_{62}$]) used as internal standards for gas chromatography (GC) analysis were obtained from Sigma–Aldrich (Milwaukee, WI, USA).

Table 1: Greases and lubricating oils and their defining characteristics

Grease/oil name	Base oil type	Thickener/additive	Dropping point (°C)	Viscosity at 100 °C cSt (ASTM D 445)	Applications
Army Grease	Hydrocarbon	Lithium complex	260+	N/A	Military lubricating grease
AeroShell #5	High viscosity mineral	Microgel® (clay)	260+	31.8	High temp. aircraft wheel bearing and engine accessory grease
AeroShell #6	Mineral	Microgel® (clay)	260+	5.5	General purpose airframe grease used in plain and anti-friction bearings
AeroShell #14	Mineral	Calcium	150	3.1	Helicopter grease, main and tail rotor bearings

AeroShell #17	Synthetic diester	Microgel* (clay)	260+	3.1	Extreme pressure multipurpose grease for heavily loaded sliding steel surfaces, bogie pivot pins on jet aircraft landing gear assemblies
AeroShell #22	Synthetic hydrocarbon	Microgel* (clay)	260+	5.8	Aircraft wheel bearings, engine accessories and airframe lubrication
Castrol 10W-30	Hydrocarbon	None	N/A	N/A	Motor engine oil (petroleum base oil)
Castrol 10W-30 [Syntec Blend]	Synthetic hydrocarbon	None	N/A	N/A	Motor oil (synthetic base oil)
Texhvi 3	Light paraffinic mineral Oil	None	204	3.31	Base Oil
Texhvi 4	Heavy paraffinic mineral Oil	None	210	4.0	Base Oil
Chrysler ATF	Mix Tex-hvi3/4	N/A	N/A	N/A	Automatic transmission oil (synthetic base oil)

Subcritical Water Treatment System Setup

The degreasing experiments were carried out with the continuous flow subcritical water extraction system shown in Fig. 1. Deionized and deoxygenated (pre-purged with helium) water was pumped through a preheat coil (stainless steel tubing of 1.6-mm o.d., 0.5-

mm i.d., and 3.5-m length) located in a temperature programmable GC oven (HP Model 5890, Hewlett Packard, Houston, TX, USA) using a high performance liquid chromatography (HPLC) pump (Constant-Flow HPLC Pump, Cole-Parmer, Vernon Hills, IL, USA). The subcritical water (or steam) was then passed through a stainless steel extraction cell (9.4-mm i.d. and 90-mm length, 6.315-mL total volume, Alltech, Deerfield, IL, USA). The cell was almost fully filled with metal ball bearings (4.75-mm o.d.), and an aliquot (~0.03 g) of grease or lubricating oil was uniformly applied on their surfaces. After the subcritical water (or steam) left from the oven, it passed through a cooling coil to drop the effluent temperature to the ambient temperature. For subcritical water operation, the system was pressurized by controlling the tubing line with a back-pressure regulator (Tescom, Elk River, MN, USA). After the extraction, the ball bearings were collected from the extraction cell. Effects of run time (1–90 min), temperature (100–250 °C), pressure (10.3–37.9 bar), and flow rate (0.5–5 mL/min) on the degreasing efficiency were assessed to find the best-optimized operating condition. For steam operation, the system was not pressurized and the back-pressure regulator was left open.

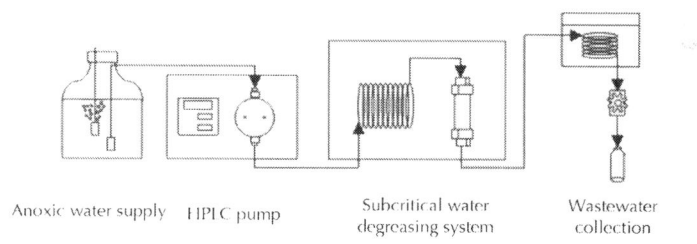

Figure 1: Schematic of superheated water degreasing system.

Degreasing with Conventional Solvents

Conventional degreasing agents (hexane, DCM, TCE, methanol, and Zep cleaner) were used to compare their degreasing efficiencies with those obtained by the subcritical water and steam degreasing systems. The reactor setup for this experiment was the same as that for the subcritical water treatment system. Each degreasing agent fluid was passed through the extraction cell at 2 mL/min for 60 min with neither heating nor pressurizing. After the extraction, the ball bearings were collected from the extraction cell and the degreasing efficiency was quantified.

Corrosion Test

The compositions of the metal ball bearings employed in the corrosion test are provided in Table 2. After the ball bearings were collected from the extraction cell, they were air-dried overnight. Physical damages formed on their surfaces, which may result in corrosion, were examined by a scanning electron microscope (SEM, Phillips XL30FEG, Philips, Amsterdam, the Netherlands).

Table 2: Chemical composition of components comprising metal ball bearings (max. wt%, primary metal fraction not included)

	C	Mn	P	S	Si	Cr	Mo	Pb	Zn
Brass								0.7	36.5
Stainless steel (440C)	1.2	1	1	0.04	0.03	18	0.75	–	–
Chrome steel (52100)	1.1	0.45	0.025	0.025	0.35	1.6	–	–	–
Carbon steel (Gr.200)	0.2	0.9	0.04	0.5	0.2	–	–	–	–

Analytical Procedures

To characterize the chemical composition and to quantify the amount of greases and lubricating oils, an analytical method was developed using horizontal attenuated total reflectance Fourier transform infrared (FTIR) spectroscopy (Thermo Nicolet Nexus 670, Thermo Scientific, Waltham, MA, USA). After each experimental run, the ball bearings were collected from the reactor cell and transferred into a 30-mL glass beaker. Then, 20 mL of hexane was added and stirred vigorously to dissolve the grease or oil remaining on the ball bearings. An aliquot (0.2 mL) of the hexane extract was then spread uniformly over an FTIR ZnSe crystal and then it was removed by evaporation. The concentration was quantified in terms of the area under the spectral signal integrated between the wave numbers of 2695 and 3190 cm^{-1} ascribed to the various C—H stretching, the wave numbers of around 1715, 1610, and 1450 cm^{-1}, and the wave numbers between 975 and 1460 cm^{-1}, assigned to the various C=O stretching of carboxyl, aldehyde, and ketone, C=C stretching of aromatic rings, and C—H deformation of long-chain aliphatic hydrocarbons such as paraffins, and C—O stretching of aryl esters, phenols, methoxy, and carbonyl groups, respectively [14]. The area was easily converted to the concentration unit by comparing each of sample spectra with those of grease or lubricating oils of no subcritical water treatment (control) as presented in Fig. SD1. Sufficient accuracy (>98%) and precision (standard error of ~3%, n = 10) were achieved from this method which was confirmed by the measurement of total organic carbon (TOC) levels for petroleum hydrocarbon standards. The TOC was determined using a TOC analyzer (Sievers 5310C, GE Analytical Instruments, Boulder, CO, USA) with potassium hydrogen phthalate as an external standard. The removal efficiency was determined using the following equation:

Removal efficiency

$$= \left[\frac{\text{concentration of lubricant reamaining}}{\text{concentration of lubricant initially applied}} \right] \times 100\% \quad (1)$$

Greases were also characterized by a GC (Agilent 7890A, Agilent Technology, Inc., Palo Alto, CA, USA) equipped with a flame ionization detector using an HP-5 column (30-m length × 0.25-mm i.d. and 0.25-μm film thickness). Aliquots of greases were solubilized in hexane and analyzed to avoid the loss of grease due to volatilization.

RESULTS

Characterization of Greases Selected for System Optimization

Out of the greases employed in this study, two aviation greases (AeroShell #6 and AeroShell #14) with dropping points and viscosities lower than other greases (Table 1) were selected to determine optimized operating parameters. The major components of greases and lubricating oils are base oils, a mixture of various heavy-molecular-weight and long-chain-saturated hydrocarbons, polyalphaolefins, alkylated aromatics, polyglycols, silicone oils, etc. [1]. Likewise, our FTIR analysis demonstrated that both aviation greases consist primarily of saturated hydrocarbons and various alkylated hydrocarbons such as ketones, esters, phenolic, carbonyl, carboxyl, and methoxy groups (Fig. SD1). The major components were found to be long-chain-aliphatic hydrocarbons ($>C_{20}$) as evidenced in Fig. SD2.

Optimization of Subcritical Water Degreasing System

The effect of run time on the degreasing efficiency was examined as a primary system operating factor, as shown in Fig. 2a. The subcritical water was supplied at a fixed flow rate of 2 mL/min and the temperature was maintained at 150 °C. As run time increased, the removal efficiencies of greases sharply improved. At maximum, 82.7% and 89.7% of AeroShell #6 and AeroShell #14 were removed, respectively, over the course of 60 min of system run. No further notable improvement was observed with increasing run time. The slightly higher degreasing rate for AeroShell #14 than that for AeroShell #6 indicates that the physical property of grease particularly represented by their dropping points was attributable to their degreasing efficiency. As noted in Fig. 2b, the temperature of the subcritical water, which greatly changes the properties of the water, had a significant impact on the removal efficiency for both greases. Removal efficiencies for AeroShell #6 and AeroShell #14 were only 29.0% and 42.3% at 100 °C, respectively, but they increased substantially with increasing temperature: there was a 1.7–1.8 fold increase in removal efficiencies with a temperature increase of 50 °C. Similar to the case of run time, the degreasing rate increase was only marginal when the temperature increased beyond 150 °C. The effect of subcritical water flow pressure on the removal efficiency was almost negligible (Fig. 2c). The flow rate was changed to examine the degreasing performance with regard to reactor dynamics. The removal efficiency of AeroShell #6 and AeroShell #14 increased markedly from 16.3% and 25.3% at 0.5 mL/min to 82.7% and 91.3% at 2 mL/min, respectively. This means that the increase of subcritical water supply (from 30 mL to 120 mL) also contributed to the increased degreasing efficiency. To this end, the best-optimized operating condition for the degreasing with subcritical water was determined to be 60 min of run time at 150 °C and 2 mL/min of flow rate, which corresponded to 120 mL of water supply.

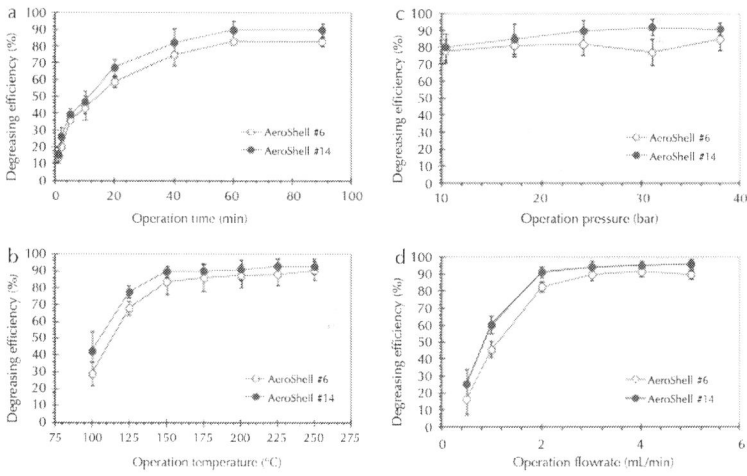

Figure 2: Effects of (a) run time, (b) temperature, (c) pressure, and (d) flow rate on the degreasing efficiency of AeroShell #14 by subcritical water

extraction system. Operating conditions except for variable factors were 60 min of run time at 150 °C, 10.3 bar, and 2 mL/min of flow rate.

Degreasing by Subcritical Water Extraction

Degreasing efficiency for various greases employed in this study was evaluated under the best-optimized operation conditions as described in the previous section. As presented in Fig. 3, AeroShell #6 and AeroShell #14 exhibited the greatest removal efficiencies (83.0% and 91.3%, respectively). However, much lower degreasing rates were obtained from other greases (40.7–55.3%). Interestingly, the degreasing efficiency was low for the greases with high viscosity and high dropping point as provided in Table 1: e.g., Army Grease, AeroShell #5, AeroShell #17, and AeroShell #22. In addition, the greases that contain clay-based thickener, Microgel® (e.g., AeroShell #5, AeroShell #17, and AeroShell #22) were relatively resistant to the subcritical water treatment.

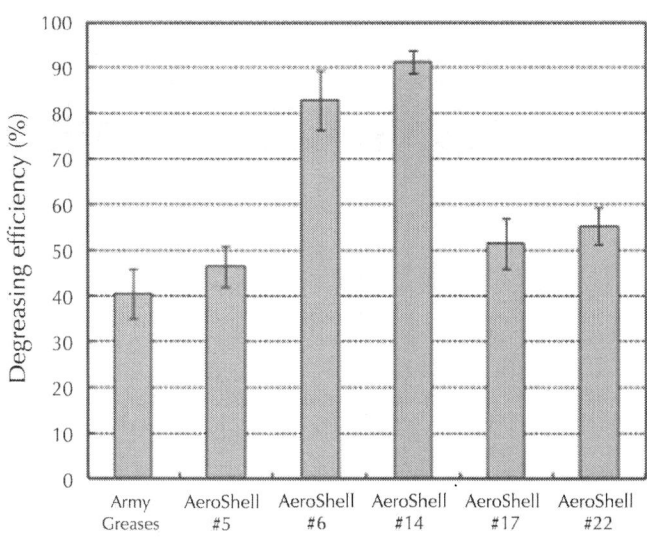

Figure 3: Degreasing efficiency by subcritical water extraction system at 150 °C and 2 mL/min for 60 min. Error bars denote standard deviation of triplicate runs.

Removal of Lubricating Oils by Subcritical Water Extraction

The removal of lubricating oils was also examined using the subcritical water extraction system with the same reactor setup as described before. The results are depicted in Fig. 4. The removal efficiencies for both synthetic and petroleum-distillate motor oils (Castrol 10W-30 and Castrol 10W-30 [Syntec Blend], respectively) were ~90%. Another experiment was conducted to test how subcritical water treatment would perform on an automatic transmission fluid (Chrysler ATF).

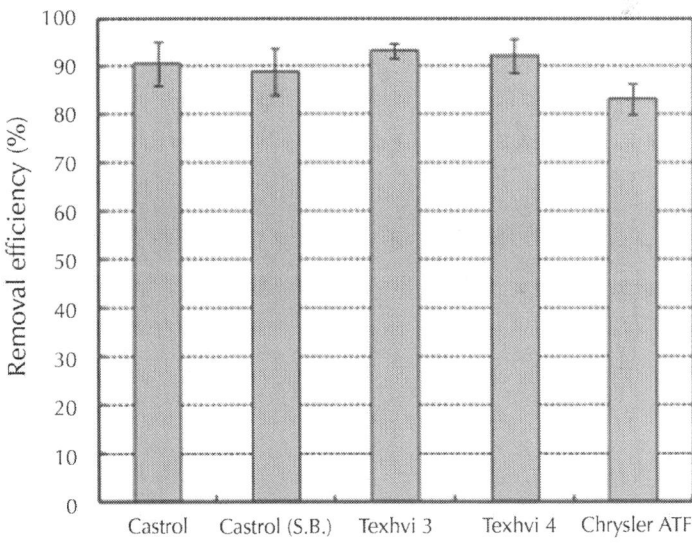

Figure 4: Removal efficiencies of lubricating oils by subcritical water extraction system at 150 °C and 2 mL/min for 60 min. Error bars denote standard deviation of triplicate runs.

This fluid was produced by the mixing of two base oils (Texhvi3 and Texhvi4) with various unknown additives to impart additional properties. Subcritical water removed both the base oils fairly efficiently (83.3–93.3%), but had little effect on the combined product. It is to be noted that Texhvi4, composed of heavy paraffins, showed slightly lower removal efficiency than Texhvi3. However, the differences in the removal efficiencies for all lubricant-base oils tested were not statistically significant, confirmed by the analysis of variance (ANOVA, p value = 0.057). These results indicated that base oils without additives such as thickeners can be removed by the subcritical water treatment at a highly efficient rate.

Degreasing with Conventional Degreasing Agents

The degreasing efficiency of subcritical water extraction for the removal of AeroShell #14 was compared with those determined using conventional cleaning agents, Zep cleaner (aqueous cleaning solvent), hexane, TCE, DCM, and methanol, including steam as shown in Fig. 5. 5 replicate runs were conducted, and the differences in the degreasing efficiencies for the methods examined were tested by ANOVA and the homoscedastic Student's t-test analyses. The ANOVA confirmed that the differences in the degreasing efficiencies of all methods tested were statistically significant (p value = 6.82×10^{-5}). The comparison of the degreasing efficiency of subcritical water extraction with that of each of the other treatment methods also revealed that the former was statistically superior (at a significance level of 5%) to the latter, supported by the Student's t-test (p value for each of the comparisons = 1.83×10^{-4}–5.17×10^{-3}). Interestingly, steam treatment efficiency was lower than that of subcritical water treatment by 15.6%, most likely due to the difference in cleaning phenomena between vapor and liquid cleaning. Vapor degreasing is carried out by condensation on the surface of the grease and then desorbing it from metal surface by solubilization, both of which can be limited by the surface area for condensation and solubilizing capacity of solvent. The physical cleaning effect (e.g., mechanical turbulence) that can be created by vapor degreasing should have been very limited.

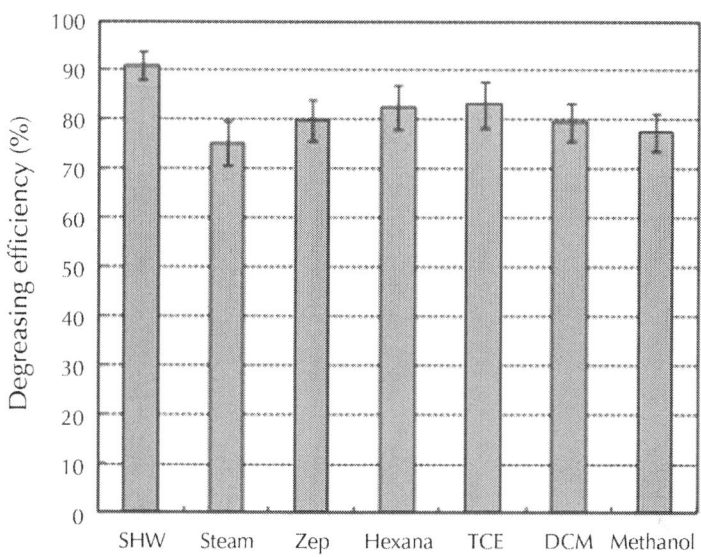

Figure 5: Comparison of degreasing efficiencies of various cleansing agents. Temperature, flow rate, and run time for all degreasing operations were 25 °C, 2 mL/min, and 60 min, respectively, except 150 °C for subcritical water and steam runs. Error bars denote standard deviation of five run times. In legend, SHW denotes superheated (subcritical) water treatment.

Corrosion Test

One of main concerns of using water at high pressures and temperatures as degreasers is the possibility of corrosion of metal parts. Thus, the surfaces of various kinds of metal ball bearings were analyzed directly by visual inspection using an SEM to examine the formation of corrosion after the subcritical water treatment. Only a small pit was found on the surface of brass ball bearings with subcritical water treatment at 150 °C (Fig. 6a). At 250 °C, the color became a little darker, and small-sized pits were observed (Fig. 6b). While the surface of brass was not very smooth to begin with, surface reactions and some minor pits were observed after the 250 °C run. Stainless steel maintained its silver color and shine after the operation at 150 °C (Fig. 6c), but there was a slight color change at 250 °C with the loss of the shiny finish, and only very small pits were found (Fig. 6d).

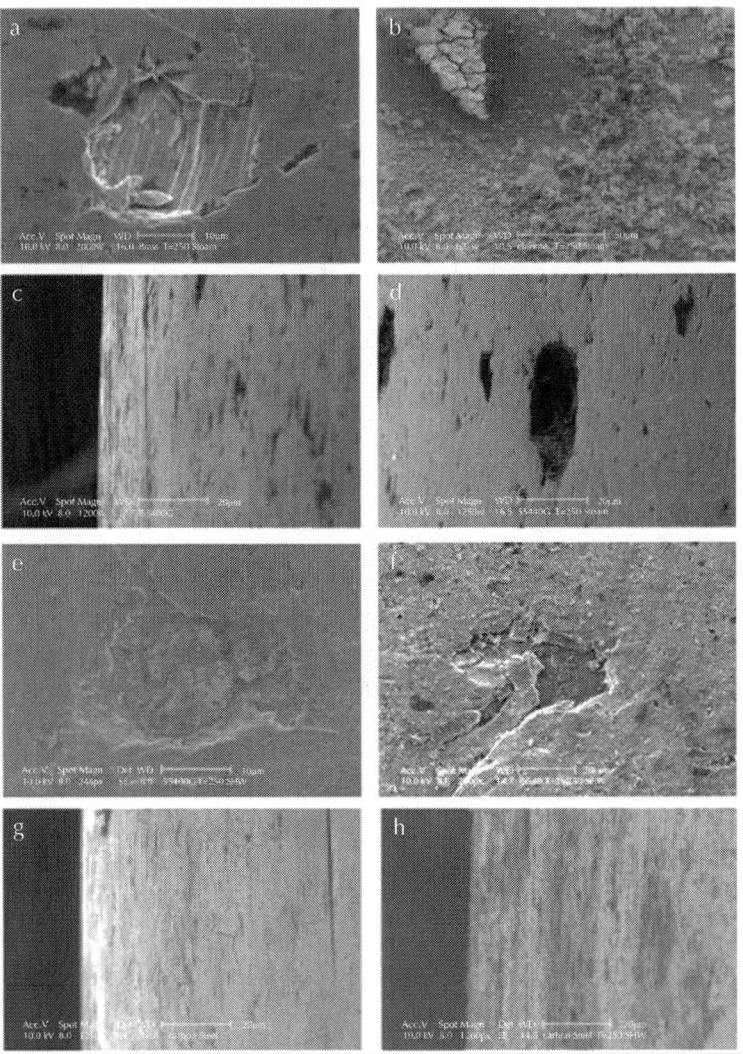

Figure 6: SEM images of the surfaces of ball bearings treated by subcritical water extraction. (a) A small pit (~20 μm diameter) was formed on the surface of a brass ball bearing after treatment at 150 °C for 60 min. (b)

Cracking and streaks were formed on the surface of a brass ball bearing by surface oxidation reaction after treatment at 250 °C for 60 min. The surface of a stainless steel ball bearing (c) before and (d) after treatment at 250 °C for 60 min; pitting has started to occur across the entire surface. A pit was found on the surface of the chrome steel ball bearing after treatment (e) at 150 °C for 90 min and (f) at 250 °C for 60 min; these also show some surface reactions, causing the surface to become rough. The surface of a carbon steel ball bearing (g) before and (h) after treatment at 250 °C for 60 min.

Chrome steel also lost its shine and changed color from silver to brown and eventually gray with the 150 °C and 250 °C treatments, respectively (Fig. 6e and f). Streaks were also found on the surface of the 150 °C run. The surface became very dark and rough after the 250 °C run, and some pitting was found. Carbon steel showed some color changes, with black spots appearing on the surface after the operation at 150 °C, but it still had a fairly smooth surface with no pitting (Fig. 6g). At 250 °C, the color changed from bluish to brown, but the shine was still apparent with a constant smoothness (Fig. 6h). Overall, the physical damages due to corrosion were almost negligible after the treatment at 150 °C, and the damages found in the case of treatment at 250 °C were also not significant. Thus, all metal parts tested were found to be corrosion-resistant to the subcritical water treatment under the anoxic condition in this study.

DISCUSSION

The properties of water change remarkably under the conditions of elevated temperatures and pressures (i.e., subcritical water) [15] and [16]. Subcritical water is capable of dissolving many organic compounds that are insoluble under ambient conditions due to the marked decrease in its ε value from 80 at 25 °C to 27 at 250 °C, which is similar to organic solvents such as acetone, ethanol, and methanol [17] and [18]. The solubilities of hydrophobic organic compounds in the superheated water increase by approximately an order of magnitude for every 50 °C increase [11]. Thus, gradually increasing the temperature of subcritical water in the washing procedure can facilitate the removal of various organic compounds [19]. A number of prior studies demonstrated that benzene, n-heptane, n-pentane, 2-methyl pentane, toluene, polycyclic aromatic hydrocarbons, polychlorinated biphenyls, chlorophenols, alkyl benzenes, anilines, and soil organic matter can be effectively extracted from environmental samples using subcritical water [20], [21],[22] and [23]. On the contrary, it was found in these studies that the performance of degreasing with subcritical water was pretty much constant in the pressure range tested. Similar to this, it has been demonstrated that the extraction of hydrophobic organic compounds, such as long aliphatic chain hydrocarbons, cannot be enhanced by a pressure increase since the pressure of superheated fluids has a minor effect on the properties of water, which is represented by ion product and ε value [24] and [25]. Also note that density of subcritical water is not markedly affected by pressure change, which may have resulted in almost no positive physical cleaning effect. In the meantime, it has been reported that the flow rate of subcritical water up to 1.1 mL/min has a positive effect on the extraction of highly hydrophobic organic compounds from soils [16] and [20]. It should be noted that as the velocity of subcritical water in the extraction cell increased (a flow rate of 1 mL/min corresponded to 1.44 cm/min of superficial velocity and a Reynolds number of 2.26), the mechanical turbulence became substantial, which may have resulted in an effective physical removal of the viscous and solid-like greases trapped in the interspaces of ball bearings. Such a fluid eddy created by the subcritical water, flowing through a tortuous porous path between ball bearings in the extraction cell, may have been attributed to the stronger hydrodynamic impacts (e.g., mixing, collision between fluids and ball bearings, and washout).

Given that the system was operated under identical conditions for each of the greases, the different cleaning efficiencies were attributed to their physical properties and compositions. AeroShell #5, AeroShell #17, and AeroShell #22 contain a clay-based thickening agent, Microgel®, which should have resisted the subcritical water extraction, resulting in lower degreasing efficiencies. Army grease contains a metal (lithium)-based thickener, which appeared to give physical resistance to the extraction as well. Another factor related to degreasing efficiencies was the viscosity of grease. The lowest degreasing efficiency was achieved from AeroShell #5, which is the most viscous and the least dissociating grease (dropping point ≥ 260 °C). On the contrary, the highest degreasing efficiency was obtained from the least viscous and the most dissociating grease (dropping point = 150 °C), which is Aeroshell #14. Some current degreasing procedures use strong acids or bases rather than organic solvents for washing the grease from part surfaces. Fortuitously, pure liquid water at 150 °C behaves as both a weak acid and a weak base, adding to the grease-removal power of subcritical water. Hot water is more acidic and basic than that at ambient temperature because its ion dissociation product increases by about 3 orders of magnitude as the temperature increases from 25 °C to 150 °C [9].

The relatively polar organic molecules of grease, such as the long chain fatty acids of the soap thickeners, should be solubilized and extracted at relatively low temperatures [10] and [16]. Optimal grease thickening is commonly achieved with soaps of fatty acids that are approximately 18 carbon atoms in chain length[26]. Fatty acids of this length should be readily solubilized in subcritical water at 150 °C. Also, metal ions of soap thickeners are readily soluble in hot liquid water. Virtually all thickener compounds that are not removed with subcritical water at lower temperatures are expected to readily dissolve as temperatures approach 250 °C. Grease additives are typically found in the metal soap thickener phase so that they will effectively compete with the soap molecules for the metal surfaces. The majority of grease additives, such as anti-oxidants, anti-wear additives, corrosion inhibitors, and dyes, are also organic molecules. Thus, the solubility of grease can increase in the subcritical water, and therefore, the grease can be effectively extracted from metal parts along with thickener phases. In addition, the subcritical water has been known to be effective at decreasing viscosity and surface tension of grease applied on the surfaces of metal parts, which is expected to improve degreasing performance markedly [7] and [27]. In the meantime, as addressed, the recovery and recycling of this environmentally benign solvent, water, must be noted as another benefit. Currently, water separators are used to recover and recycle organic solvents used in degreasing [8]. While this achieves some reduction of solvent waste, the affinity of these solvents for greases allows them to be reused for only a few cycles before they lose their effectiveness.

Corrosion is controlled by a variety of factors including the dissociation of acids and the solubility of oxygen, hydrogen, protecting oxide groups, and corrosion products. As water temperature increases, its density and ε value decrease, with its ion product reaching a maximum around 300 °C. This allows increased solubility of gases such as oxygen, which can cause corrosion even at low concentrations. The oxide layer, which protects almost all metals, can also be destroyed by high temperature and pressure water by creating more thermodynamically favorable species and a less kinetically controlled environment. The rate of corrosion is mainly determined by the dissolution of the non-protecting corrosion products formed on the surface [9] and [28]. While high corrosion rates have been found under supercritical conditions, relatively low corrosion rates have been observed at subcritical temperatures [29] and [30]. Thus, the subcritical water, in particular, under the anoxic condition has an advantage in terms of minimizing the possibility of corrosion as demonstrated in this study.

CONCLUSIONS

In summary, an environmentally benign solvent, water in the form of subcritical fluid, was used as a degreasing solvent in this study. It was found that its degreasing efficiency was highly dependent on temperature, but independent of pressure of subcritical flow. The optimized operation conditions for degreasing using subcritical water extraction was determined as follows: >60 min of run time at >150 °C and 2 mL/min of flow rate. The degreasing efficiency was found to be sensitive to the physical characteristics and chemical compositions of grease. Greases with a high dropping point and high viscosity were resistant to superheated water treatment, resulting in low degreasing efficiency. Base oils (lubricating oils), a major component of grease, were effectively removed by subcritical water extraction (degreasing rate of 83.3–93.3%). The subcritical water treatment demonstrated a better degreasing efficiency than conventional degreasing methods that employ toxic organic or caustic aqueous-based solvents. Only minor physical damages were observed on the metal parts after the subcritical water treatment, indicating that all metal parts tested are corrosion-resistant. The results of this study supported the idea that subcritical water treatment can be used as an environmentally friendly replacement for conventional degreasing methods.

ACKNOWLEDGMENTS

We appreciate Carl Lenker for his laboratory work and data analysis. Funding for this research was provided by the National Center for Environmental Research under the US Environmental Protection Agency (Grant R-828246), and partial supports were provided by the Geo-Advanced Innovative Action Projects(2012000550002 and 2012000550008) and the Waste to Energy and Recycling Human Resource Development Project (YL-WE-12-001) funded by the Ministry of Environment and the U-City Master and Doctor Course Grant Program funded by the Korea Ministry of Land, Transport and Maritime Affairs.

REFERENCES

1. R.M. Mortier, S.T. Orszulik, Chemistry, Technology of Lubricants, VCH Publishers, Inc., New York, 1992.
2. H.B. Silver, I.R. Stanley, The effect of the thickener on the efficiency of loadcarrying additives in greases, Tribology 7 (1974) 113–118.

3. USEPA, Guide to Cleaner Technologies: Cleaning and Degreasing Process Changes, Office of Research and Development, USEPA, Washington, DC, 1994 (EPA/625/R-93/017).
4. M. Menta, J. Frayret, C. Gleyzes, A. Castetbon, M. Potin-Gautier, Development of an analytical method to monitor industrial degreasing and rinsing baths, J. Cleaner Production 20 (2012) 161–169.
5. J.A. Mertens, Vapor degreasing with chlorinated solvents, Metal Finishing 108 (2010) 23–32.
6. K. Hartonen, K. Inkala, M. Kangas, M.-L. Riekkola, Extraction of polychlorinated biphenyls with water under subcritical conditions, J. Chromatography 785 (1997) 219–226.
7. R.M. Smith, Extractions with subcritical water, J. Chromatography A 975 (2002) 31–46.
8. M. Finkbeiner, E. Hoffmann, G. Kreisel, Environmental auditing: the functional unit in the life cycle inventory analysis of degreasing processes in the metal processing industry, Environmental Management 21 (1997) 635–642.
9. G.C. Akerlof, H.I. Oshry, The dielectric constant of water at high temperatures and in equilibrium with its vapor, J. American Chemical Society 72 (1950) 2844–2847.
10. A.G. Carr, R. Mammucari, N.R. Foster, A review of subcritical water as a solvent and its utilisation for the processing of hydrophobic organic compounds, Chemical Engineering J. 172 (2011) 1–17.
11. D.J. Miller, S.B. Hawthorne, Solubility of polycyclic aromatic hydrocarbons in subcritical water from 298K to 498K, J. Chemical and Engineering Data 43 (1998) 1043–1047.
12. D.J. Miller, S.B. Hawthorne, Solubility of liquid organics of environmental interest in subcritical (hot/liquid) water from 298K to 473K, J. Chemical and Engineering Data 45 (2000) 78–81.
13. A. Kubátová, D.J. Miler, S.B. Hawthorne, Comparison of subcritical water and organic solvents for extracting kava

lactones from kava root, J. Chromatography A 923 (2001) 187–194.
14. B.C. Smith, Fundamentals of Fourier Transform Infrared Spectroscopy, 2nd ed., CRC Press, Boca Raton, 2011.
15. S. Hashimoto, K. Watanabe, K. Nose, M. Morita, Remediation of soil contaminated with dioxins by subcritical water extraction, Chemosphere 54 (2004) 89–96.
16. M.Z. Ozel, F. Gogus, A.C. Lewis, Subcritical water extraction of essential oils from Thymbra spicata, Food Chemistry 82 (2003) 381–386.
17. B. Aliakbarian, A. Fathi, P. Perego, F. Dehghani, Extraction of antioxidants from winery wastes using subcritical water, J. Supercritical Fluids 65 (2012) 18–24.
18. B. Kuhlmann, E.M. Arnett, M. Siskin, Classical organic reactions in pure subcritical water, J. Organic Chemistry 59 (1994) 3098–3101.
19. M. Siskin, A.R. Katritzky, A review of the reactivity of organic compounds with oxygen-containing functionality in supercritical water, J. Analytical and Applied Pyrolysis 54 (2000) 193–214.
20. S.B. Hawthorne, Y. Yang, D.J. Miller, Extraction of organic pollutants from environmental solids with sub- and supercritical water, Analytical Chemistry 66 (1994) 2912–2920.
21. A.E. Latawiec, B.J. Reid, Sequential extraction of polycyclic aromatic hydrocarbons using subcritical water, Chemosphere 78 (2010) 1042–1048.
22. E. Priego-López, M.D.L. Castro, Subcritical water extraction of linear alquilbenzene sulfonates from sediments with on-line preconcentration/derivatization/detection, Analytica Chimica Acta 511 (2004) 249–254.
23. Y. Yang, S.B. Hawhtorne, D.J. Miller, Class-selective extraction of polar, moderately polar, and nonpolar organics from hydrocarbon wastes using subcritical water, Environmental Science and Technology 31 (1997) 430–437.

24. M.S. Chang, J.Y. Shen, S.-H. Yang, G.J. Wu, Subcritical water extraction for the remediation of phthalate ester-contaminated soil, J. Hazardous Materials 192 (2011) 1203–1209.
25. J. Rincón, P. Canizares, ˇM.T. García, Regeneration of used lubricant oil by ethane extraction, J. Supercritical Fluids 39 (2007) 315–322.
26. D.Klamann, Lubricants and Related Products,Verlag Chemie GmbH,Weinheim, 1984.
27. R.M. Smith, Subcritical water chromatography – a green technology for the future, J. Chromatography A 1184 (2008) 441–445.
28. P. Kritzer, N. Boukis, E. Dinjus, Factors controlling corrosion in hightemperature aqueous solutions: a contribution to the dissociation and solubility data influencing corrosion processes, J. Supercritical Fluids 15 (1999) 205–227.
29. L.B. Kriksunov, D.D. Macdonald, Corrosion in supercritical water oxidation systems: a phenomenological analysis, J. Electrochemistry Society 142 (1995) 4069–4073.
30. P.Kritzer, N. Boukis, E. Dinjus, Corrosion of alloy 625 in high temperature sulfate solutions, Corrosion 54 (1998) 689–699.

Chapter 5

The Intricate Structural Chemistry of Base Excision Repair Machinery: Implications for DNA Damage Recognition, Removal, and Repair

Kenichi Hitomi[a,b,c], Shigenori Iwai[a], and John A. Tainer[b,c]

[a]Division of Chemistry, Graduate School of Engineering Science, Osaka University, 1–3 Machikaneyama, Toyonaka, Osaka 560-8531, Japan

[b]Department of Molecular Biology and the Skaggs Institute for Chemical Biology, The Scripps Research Institute, 10550 North Torrey Pines Road, La Jolla, CA 92037, USA

[c]Life Sciences Division, Lawrence Berkeley National Laboratory, 1 Cyclotron Road, Mail Stop 74R157, Berkeley, CA 94720, USA

ABSTRACT

Three-dimensional structures of DNA N-glycosylases and N-glycosylase/apyrimidine/apurine (AP)-lyase enzymes and other critical components of base excision repair (BER) machinery including structure-specific nuclease, repair polymerase, DNA ligase, and PCNA tethering complexes reveal the overall unity of the simple cut and patch process of DNA repair for damaged bases. In general, the damage-specific excision is initiated by structurally-variable DNA glycosylases targeted to distinct base lesions. This committed excision step is followed by a subsequent damage-general processing of the resulting abasic sites and 3' termini, the insertion of the correct base by a repair DNA polymerase, and finally sealing the nicked backbone by DNA ligase. However, recent structures of protein–DNA and protein–protein complexes and other BER machinery are providing a more in-depth look into the intricate functional diversity and complexity of maintaining genomic integrity despite very high levels of constant DNA base damage from endogenous as well as environmental agents. Here we focus on key discoveries concerning BER structural biology that speak to better understanding the damage recognition, reaction mechanisms, conformational controls, coordinated handoffs, and biological activities including links to cancer. As the three-dimensional crystal and NMR structures for the protein and DNA complexes of all major components of the BER system have now been determined, we provide here a relatively complete description of the key complexes starting from DNA base damage detection and excision to the final ligation process. As our understanding of BER structural biology and the molecular basis for cancer improve, we predict that there will be multiple links joining BER enzyme mutations and cancer predispositions, such as now seen for MYH. Currently, structural results are realizing the promise that high-resolution structures provide detailed insights into how these BER proteins function to specifically recognize, remove, and repair DNA base damage without the release of toxic and mutagenic intermediates.

INTRODUCTION

In the face of constant and high levels of DNA base damage, genetic integrity is protected by robust and dynamic base excision repair (BER) systems in cell. Compared with other DNA repair machinery, such as nucleotide excision repair, base excision repair systems are vertically well conserved, from bacteria to humans in terms of the core components of BER machinery (Fig. 1) [1], [2], [3], [4] and [5]. In cells, however, BER acts in complex protein networks that vary among different types of organisms and that are not independent of other DNA metabolic regulations including replication and transcription. In this sense, BER systems in Escherichia coli are primarily a paradigm for DNA repair in E. coli, not in humans, because the network of connections can vary despite the commonality of overall pathway steps and the apparent similarities among core catalytic domains. Yet, all living organisms share the DNA code and chemistry and consequently similar BER systems to preserve genetic integrity in the face of common types of base damage. So we benefit from the expanding number of detailed three-dimensional structures for BER components and dynamic machinery to reveal both the diversity and the universality in the responses to DNA base damage for life to exist [6].

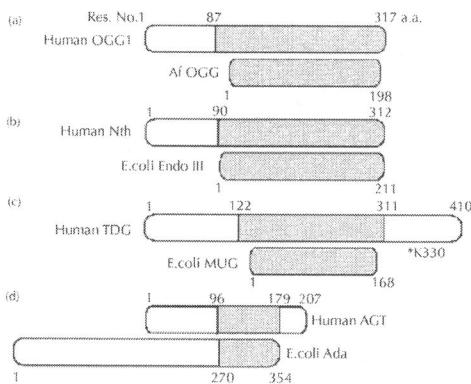

Figure 1: Primary structural comparisons between bacterial and human

enzymes: (a) the HhH-type glycosylases pyrimidine type, E. coliEndoIII and human Nth; (b) purine family, AfOGG and hOGG1; (c) UDG family, E. coli MUG and human TDG (asterisk shows the unique SUMOlyation site); (d) O^6-methylguanine DNA methyltransferase, E. coli Ada and human AGT. Conserved catalytic regions are highlighted in gray.

BER occurs in two overall steps: first, damage-specific recognition and excision carried out by glycosylases targeted to distinct base lesions, and second, a damage-general stage during which the resulting central apyrimidine/apurine (AP) site intermediates and 3' termini are processed by structure-specific nucleases such as FEN-1, followed by DNA repair synthesis and ligation [4] and [7]. While diverse types of DNA base damage occur in cells, all organisms have evolved efficient, specific means to find, remove, and repair such base damage [1]. Furthermore, as base damage is the most common form of DNA damage in cells, BER is one of the major pathways for dealing with most DNA damage [3] and [5].

The damage recognition of BER relies on distinct glycosylases; these are divided into simple N-glycosylases and bifunctional N-glycosylase/AP-lyase enzymes [2]. In the case of pure N-glycosylases, the specifically recognized damaged base is first removed, then the abasic site is acted upon by a separate enzyme with AP-lyase activity, such as APE1, which cleaves the DNA backbone. Even the removal of normal bases by DNA glycosylases is potentially mutagenic or lethal if the resulting AP site is improperly repaired [8], [9] and [10]. N-glycosylase/AP-lyase enzymes combine the first two functional steps within a single enzyme [2], [4] and [7]. Remarkably, most glycosylases share a fold with common motif of the active site. The damage-general stages of BER require a repair DNA polymerase to insert the proper base, and polymerase (pol) plays a major repair role in a complex with XRCC1. Finally, the DNA strand is rejoined by a DNA ligase, ligase III, for complete repair of the damaged base site. In addition to the so-called short-patch repair path, glycosylase products are alternatively processed by long-patch repair involving polymerase, FEN-1, PCNA, RFC, and ligase I [2] and [7].

Environmental agents such as UV light, ionizing radiation, and alkylation of DNA create distinct DNA base lesions including deamination, depurination, and depyrimidation [11]. The majority of DNA damage is caused by endogenous metabolic agents, such as oxidation [4] and [7]. The BER system efficiently detects such base damage in the context of an enormous background of normal DNA bases, removes this damage, and restores the normal base to maintain genome integrity. The initial damage recognition process evidently mainly relies upon the glycosylases of BER machinery. The structural biochemistry of these enzymes has thus revealed paradigms from glycosylase structural biology for damaged base recognition. For example, similar folds, common motifs, the gylcosylase exchange of DNA base:base pairing to DNA base:protein pairing [12], a universal nucleotide flipping-out mechanism, as first discovered for the uracil-DNA gylcosylase:DNA complex [13], the helix-hairpin-helix (HhH) motif and invariant Asp residue common to the active sites of many monofunctional and bifunctional DNA glycosylases [2], and so forth [recently reviewed in ref. 6].

On the other hand, emerging structural biology has also raised deeper questions and issues. The protein fold, active channel shape and conserved motifs belie the diversity in the substrate recognition and the real difference in specific reactions, such as seen between the pure or monofunctional N-glycosylases and the bifunctional N-glycosylase/AP-lyase enzymes. Compared with the BER core enzymes of bacteria, many glycosylase counterparts in eukaryotes have added termini that elaborate the activities and functions of these enzymes in more complex organisms in ways as yet incompletely understood (Fig. 1). Similarly, the true significance of seemingly energy-consuming redundancy among BER enzyme activities remains unclear. However, the elucidation of the relevant protein–DNA and protein–protein interactions for these BER components and their interactions is providing deeper insights and a better understanding of the BER machinery in diverse cells. Such insights have biological and medical significance including assessment of public health risks from environmental toxins, as experiments are suggesting that inhibition of DNA repair may be far worse than DNA damage [14]. Here, we focus upon recent discoveries concerning the detailed structural biology of BER machinery by x-ray crystallography and NMR, especially protein–DNA and protein–protein complexes that elucidate the structural chemistry of short and long-patch BER processes. We also include enzyme–DNA complexes for proteins that perform direct damage reversal. In many ways the structural biochemistry of these damage reversal proteins in complex with DNA is uncovering unifying themes and enlightening diversity in damaged base recognition and repair. As the three-dimensional crystal and NMR structures for the protein and DNA complexes that are representatives of all major components of the BER system have now been determined, we can provide here the first relatively complete description of the key complexes starting from DNA base damage detection and excision to the final ligation process. This detailed structural information is revealing the diverse chemical anatomy for BER machinery that underscores the biologically important variations in genome maintenance among different cells and diverse organisms with implications for cancer risks, predispositions and interventions.

THE PURE OR MONOFUNCTIONAL DNA GLYCOSYLASE IN COMPLEX WITH DNA

DNA N-glycosylases hydrolyse the N-glycosylic bond between the target base and deoxyribose, thus releasing a free base and leaving an AP site in DNA [2]. Such AP sites are cytotoxic and mutagenic, and must be further processed (see below). Some DNA glycosylases also have an associated AP-lyase activity that cleaves the phosphodiester bond 3' to the AP site and are herein termed bifunctional glycosylases and considered in a separate section below. The catalytic mechanism of the pure or monofunctional glycosylases considered in this section does not involve covalent intermediates. Instead the conserved Asp residue may activate a water molecule, which acts as the attacking nucleophile. Three-dimensional structure determinations have allowed BER glycosylases to be classified into several major structural families by architectural folds [2]: helix-hairpin-helix [15], helix-two-turn-helix (H2TH) [16], and uracil DNA glycosylases (UDGs) [6]. Glycosylase structures furthermore can show a variety of additional functional domains, such as an [4Fe-4S] iron sulfur cluster, as first structurally characterized in EndoIII/Nth [17] and also found in MutY and MIG [18] and [19], a sheet in AlkA and OGG1 [20] and [21], a MutT-like domain in MutY [22], and a methyl-CpG binding domain in MBD4 [23]. It is worthwhile considering this structural variation in terms of glycosylase activity and function in cells.

The human enzyme methyl-CpG binding domain protein 4 (MBD4) uniquely contains a C-terminal glycosylase domain, which is similar to the M. thermoformicicum thymine DNA glycosylase MIG of the HhH glycosylase family, linked to an N-terminal methyl-CpG binding domain [19] and [23]. MIG recognizes G:T mispairs and removes the thymine base [19], whereas MBD4 preferentially binds 5-methyl-CpG:GpT mismatches, but will also excise uracil from G:U mispairs [24]. In the proposed mechanism for G:T mismatch recognition and N-glycosylic bond cleavage, which is consistent with structural biochemistry and site-directed mutants, MIG bond cleavage is enhanced by a physical distortion of the nucleotide that imparts a ~90° twist to the thymine base, away from its normal anti position in DNA [19], similar to the model proposed for UDG [25]. The DNA-binding model for thymine DNA glycosylases is consistent with known HhH type glycosylase–DNA complex structures, and with most other DNA repair glycosylase structures, in that binding is mediated exclusively via the DNA minor groove [2] and [13].

NMR structures have elucidated that the methyl-CpG binding domain consists of a compact α/β fold with an extended loop between two anti-parallel β strands [26] and [27], and that these β strands insert into the DNA major groove where particular residues bind methyl-CpG sequences [26]. The methyl-CpG binding domain›s DNA major-groove binding complements to the DNA minor-groove binding of the glycosylase domain, suggesting that the N-terminal methyl-CpG and C-terminal glycosylase domains of MBD4 would be able to access together 5-methyl-CpG:GpT mismatches to hold the DNA. Although the isolated glycosylase domain of MBD4 is enzymatically active [28], this dual-binding model provides a mechanism to restrict MBD4 thymine DNA glycosylase activity in vivo to biologically relevant substrates at the promoters of silenced genes without interference from the G:U mispairs that arise continuously throughout the genome from general cytosine deamination.

In general, the discoveries of common overall folds and functional motifs do not provide mechanistic details regarding the distinct specificity and often differential biological function of these enzymes. The breakthroughs concerning these issues depend upon combing genetics and biological characterizations with systematic characterizations of enzyme complexes with DNA as first shown for UDG [29]. Thus, we expect that an improved understanding of the glycosylase and methyl-CpG binding domain coordination will require detailed and systematic structural and biochemical analyses, particularly of the enzyme–DNA complexes.

The adenine DNA glycosylase, MutY, which belongs to the HhH glycosylase family [15], recognizes 8-oxoG:A mispairs and excises the adenine to avoid the formation of mutations [4], [7], [30] and [31]. MutY and its homologues have a catalytic core domain with an [4Fe-4S] iron sulfur cluster in the N-terminus [18], followed by an additional C-terminal MutT-like domain [22] and [30]. MutT catalyzes the hydrolysis of mutagenic nucleoside triphosphates, such as 8-oxoG deoxyribonucleotide triphosphates, by substitution at the rarely attacked β-P, to yield deoxyribonucleotide monophosphates and inorganic pyrophosphate, thus preventing misincorporation of this base into DNA [32]. NMR studies have shown that MutT is a mixed α/β protein with two helices and two β sheets composed of five β strands and that the C-terminal domain of MutY has similar secondary structure and topology with the MutT, despite low sequence identity between the two proteins [22] and [32]. The additional C-terminal domain of MutY also seemed to be responsible for 8-oxoG recognition, as the truncation of the domain results in loss of discrimination between 8-oxoG:A and G:A mispairs [33] and [34].

The recent full-length structure of MutY cross-linked to DNA containing an 8-oxoG:A base pair has helped to address key questions regarding how the protein recognizes both 8-oxoG and adenine bases (Fig. 2A) [30]. The catalytic core domain of MutY, including the structural [4Fe-4S] cluster, bears a striking overall similarity to Endonuclease III that excises oxidized pyrimidines [17], [18] and [35], while the C-terminal domain structure of MutY is highly similar to MutT [22] and [32]. The catalytic core and MutT-like domains both encircle the DNA, individually making close contacts to the appropriate DNA strand. Whereas the adenine is flipped out into a deep pocket similarly to other HhH type glycosylase–DNA complex structures, the 8-oxoG is not flipped out of the base stack. Although the MutT-like domain establishes extensive contacts with 8-oxoG, the interactions are quite different from those of MutT [36]. The anti-8-oxoG is stabilized in the MutY complex structure rather than the syn conformation, which is normally the energetically favored conformer when mispaired opposite adenine. The 8-oxoG is specifically recognized through a hydrogen bond to N7 from a serine hydroxyl group, which is in turn oriented by a hydrogen bond to a hydroxyl of tyrosine. The carbonyl group at C8 position may also form a hydrogen bond to the backbone amino group of the same serine, although the geometry is not ideal.

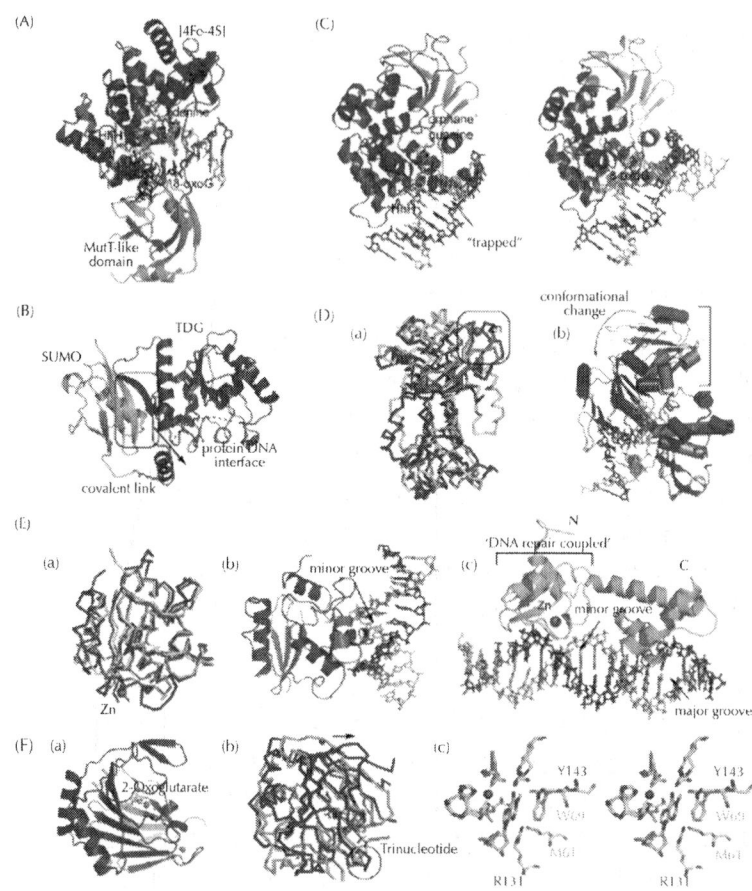

Figure 2: Structures of damage recognition proteins in complex with substrate DNA. (A) Structure of MutY complex with DNA carrying 8-oxoG:A mispair shows function for each of the respective domains of the single peptide; N-glycosylase domain, blue; MutT-like domain, green. The HhH motif is highlighted in red. (B) SUMOlylated TDG structure; TDG core domain shown in blue; accessorized region shown in green. (C) Trapped OGG discriminates against 8-oxoG (right) from normal guanine (left) in crystals. The HhH motif and N-terminal β sheets are highlighted in red and green, respectively. (D) (a) NEIL structure (yellow) overlaid with Endo VIII (blue) shows the structural conservation: zinc finger domain colored pink; "zincless" finger domain shown in orange. (b) Apo (purple) and DNA-bound (blue) forms of Endo VIII suggest that this enzyme undergoes a drastic conformational change. (E) (a) Structural conservation of

methyltransferase domain of Ada/AGT enzymes; human is shown in blue; bacterial is shown in yellow. (b) Human AGT uniquely binds the minor groove of the DNA duplex. The HTH is shown in red. (c) Two components of the E. Coli Ada N-terminal domain specifically bind to their consensus sequence. (F) (a) Human AlkB homologue, ABH3 coordinates iron and 2-oxoglutarate as cofactors. (b) Structural comparison between bacterial and human enzymes shows significant diversity in the nucleoside-binding interface. (c) Human enzyme lacks the key tryptophan.

The human homolog of MutY, MYH, is implicated in familial adenomatous polyposis, a type of hereditary colon cancer. The disease was found in one family to result from germline mutations in MYH that caused a defect in repair leading to a number of somatic mutations in the APC gene, normally linked to familial adenomatous polyposis [37]. Notably, the residue in humans corresponding to the tyrosine in the structure is mutated in some familial adenomatous polyposis patients [38]. This direct connection of the BER glycosylase to cancer resulted in part from having a good means to define and test a relevant at risk group. These results on MYH therefore underscore the important synergism between understanding the basic properties of BER enzymes and understanding how these genome guardians contribute to the prevention of human cancer. We propose that as our understanding of BER structural biology and the molecular basis for cancer continues to improve, we will increasingly discover many links joining BER enzyme mutations and cancer predispositions and that MYH is only the very beginning of such connections.

THE COVALENTLY MODIFIED GLYCOSYLASE: SUMOLYATED THYMINE DNA GLYCOSYLASE

Human TDG, a thymine DNA glycosylase was initially discovered for its ability to hydrolyze the N-glycosidic bond of the thymidine moiety in G:T mismatch [39]. TDG was later shown to remove thymine from C:T and T:T mismatches, but much less efficiently.

More importantly, it removes uracil from G:U mispair with ~10-fold higher activity than thymine, suggesting that the enzyme functions in the restoration of G:C base pairs at sites of cytosine or 5-methylcytosine deamination [40]. E. coli MUG, mismatch uracil-DNA glycosylase is closely related to human TDG (37% sequence identity in the catalytic core domain of human TDG), and they possess the same fold, whereas the bacterial homolog only removes uracil and not thymine mismatched with guanine [41]. The crystal structures of MUG and TDG have revealed an overall resemblance to uracil DNA glycosylases UDGs [12], [13], [42], [43] and [44]. Two highly conserved motifs in UDG have topological and conformational equivalents in MUG/TDG enzymes: the water activating loop (GQDPY) and the minor groove-intercalating loop (HPSPLS). The corresponding motifs in MUG are GINPGL (identical in human TDG) and NPSGLS (MPSSSS in human TDG), respectively. The latter motif forms specific hydrogen bonds with the unpaired guanine and may constitute the basis of the mismatch specificity [44]. The catalytic residues of UDG underlined in the motif are in both cases replaced by asparagines. The aspartate in the first motif in UDG activates a water molecule for nucleophilic attack on the Cl' of the deoxyribose [13]. The asparagine in MUG cannot activate the nucleophilic water, but a water molecule is found in almost the same position as seen in UDG. The tyrosine residue that provides the barrier against thymine in UDG is replaced by a glycine in MUG. The preference for G:U over G:T in MUG is likely conferred by the position of a serine hydroxyl group, which would clash with the 5' methyl group of thymine. In human TDG, the residue corresponding to the serine of MUG is substituted to an alanine. The smaller alanine side chain would allow better accommodation of thymine, explaining the expanded specificity of TDG for that base. The MUG/TDG specificity for G:U or G:T mispairs over normal G:C base pairs results not from the recognition of the scissile base itself, as in the UDG enzymes, but rather from a combination of the ease of flipping-out a base from an unstable pair over flipping from a Watson-Crick G:C pair and from the deformability of DNA at non-canonical base pairs.

As with many DNA glycosylases, the rate-limiting step in the action of TDG and MUG is its dissociation process, suggesting that the enzyme protects the harmful AP site until AP endonuclease follows the subsequent step in the BER pathway. The crystal structure of MUG in a complex with DNA shows that the enzyme binds stably to DNA by forming hydrogen bonds with the orphan guanine opposite the AP site [44]. Interestingly, the release of the DNA product can be facilitated by covalent modification as seen for human TDG, which interacts with and is covalently attached by the ubiquitin-like proteins SUMO-1 and SUMO 2/3 [45] and [46]. TDG has a lysine in the less conserved C-terminus, the ε-amino group of which can be linked with C-terminus of SUMO via an isopeptide bond. During DNA mismatch repair, the SUMO conjugation of TDG promotes the release of TDG from AP site created after base excision, and coordinates its transfer to AP endonuclease, which catalyses the further step in the repair pathway. The crystal structure of the catalytic core domain of human TDG conjugated to SUMO-1 implicates a molecular mechanism of SUMO modification mediated modulation of enzymatic properties of TDG (Fig. 2B) [47].

In the SUMO-1–TDG complex, TDG bears a striking overall similarity to MUG and the UDG family, while the SUMO-1 moiety is nearly identical to that in unconjugated SUMO-1. Significantly, the establishment of a β sheet consisting of two β strands, respectively, derived from TDG and SUMO-1 enables the C-terminal helix of TDG to protrude. The β strand of TDG contains a sequence VQEV that is similar to the consensus SUMO binding motif, V/I-X-V/I-V/I [48], critical for the TDG–SUMO-1 interaction. The helix is formed by the covalent and non-covalent contacts between TDG and SUMO-1 and sticks out from the protein surface. The resulting protruding helix presumably perturbs the protein–DNA interface and thus promotes the dissociation of TDG from the DNA molecule. The non-covalent contacts including a pivotal arginine are also essential for release from the product DNA.

THE BIFUNCTIONAL N-GLYCOSYL-ASE/AP-LYASE

Bifunctional DNA glycosylases are N-glycosylases that not only hydrolyse the N-glycosylic bond between the target base and deoxyribose but furthermore have an associated AP-lyase activity that cleaves the phosphodiester bond 3' to the AP site. Human 8-oxoguanine DNA glycosylase (hOGG1) is prototypic of the bifunctional N-glycosylase/AP-lyase enzymes that have been extensively studied (Fig. 2C) [21], [49], [50], [51], [52], [53] and [54]. Several distinct enzymes recognize 8-oxoG in different contexts because it is a major mutagenic base lesion: 8-oxoG pairs in syn conformation with adenine rather than in anti with cytosine and thereby causing G:C to T:A transversion mutations upon replication.

Human OGG excises 8-oxoG from an 8-oxoG:C pair, and mutations in hOGG1 have been linked to lung cancer in humans [55] and [56]. The overall structure contains a fold that is common to members of the HhH type glycosylase family of BER proteins, such as Endo III, AlkA, MIG, and MutY [17], [18], [19] and [20]. The N-terminal domain typically has four α helices, and the C-terminal domain has six to seven α helices. The hallmark of these proteins is a helix-hairpin-helix structural element, followed by a Gly/Pro-rich loop and a highly conserved aspartic acid, as originally discovered for Endo III. The hairpin loop shows strong sequence conservation, with consensus sequence L/F-P/K/H-G-V/I-G-K/R/T [15]. The structure of hOGG1 contains, in addition to the two-helical domains common to all known family members, a third anti-parallel-sheet domain found thus far in only the alkylation repair DNA glycosylase AlkA[20] and [21]. The function of this extended domain remains unclear.

The structure of a catalytically inactive hOGG1 enzyme core bound to 8-oxoG-containing duplex DNA revealed that hOGG1 contains an HhH motif and flips the 8-oxoG base out of the double helix into a specific recognition pocket (Fig. 2C) [21]. Human

OGG1 evidently discriminates 8-oxoG from G by using a single hydrogen bond between the Gly carbonyl and the purine N7, which is protonated only in 8-oxoG, and no direct contacts are made to the carbonyl group at C8 position. The base opposite 8-oxoG is preferentially C > T > G ≫ A, and the preference for C was shown to result from specific hydrogen bonds donated by two arginines to N3 and O_2 of the cytosine base. Although hOGG1 and AlkA have the same overall fold and share a common motif, AlkA is a simple N-glycosylase whereas hOGG1 is a bifunctional N-glycosylase/AP-lyase [20] and [21].

How bifunctional glycosylases, as represented by hOGG1, accomplish catalyses encompassing no fewer than five sequential reaction steps within a single active site has been enigmatic. A lysine has been proposed to be the key for the catalytic reaction and to attack C1' and promote -elimination [21] and [51]. Substitution of the lysine enormously decreases both N-glycosylase and AP-lyase activities and retains intact 8-oxoG in the crystal. Also the monofunctional glycosylase AlkA lacks such a lysine residue [20], and the chemical role of lysine in bifunctional glycosylases is confirmed by trapping the catalytic intermediate with $NaHB_4$ [51]. The reaction proceeds via the Schiff base and reduction by the agent gives a stable complex of the protein covalently attached to DNA. In the trapped complex structure, the cleaved 8-oxoG moiety is observed in the active site, prompting the authors to propose that 8-oxoG base itself functions in acid/base chemistry. On the one hand, however, the active site structure of the abasic-trapped complex, i.e. resultant with DNA carrying the AP site, is also nearly identical to that of the 8-oxoG-containing structure. An independent experiment reports that hOGG1 AP-lyase activity is inhibited with free 8-oxoG [57].

Endo III is also an N-glycosylase/AP-lyase. The same strategy employed to determine the hOGG1 complex was therefore also applied to an Endo III-DNA complex [35]. In the trapped Endo III structure, the excised base moiety is not observed. Notably, a structure of apo hOGG1 showed that, in the absence of DNA, the same overall enzyme conformation is conserved, but key catalytic residues, such as the lysine, are positioned improperly for catalysis [49]. Binding of the correct substrate is therefore proposed to correlate to reorientation of these side chains and subsequent catalysis. Interestingly, a hOGG1 mutant and DNA carrying 8-oxoG can be crystallized after in situ repair. The complete DNA processing in the crystallization experiment eventually generates 2′, 3′-didehydro-2′,3′-dideoxyribose as a product in a crystal, yet many biochemical data suggest that a 3′–α/β-unsaturated aldehyde is the β-elimination product from N-glycosylase/AP-lyase enzymes [52].

DNA repair systems must find damage in the context of genomic DNA. A major question for BER glycosylases concerns how damaged bases are efficiently located in this enormous background. Most base damage recognized by BER glycosylases is subtle, differing by only two atoms from a normal base. The structure of the hOGG1 and B-DNA complex proposes one possibility for efficient searching systems (Fig. 2C) [53]. A cross-link strategy with substitution of asparagine to cystine allowed hOGG1–B-DNA to be crystallized. In the hOGG complex with DNA carrying 8-oxoG, the side chain carbonyl of the asparagine makes a hydrogen bond with the exocyclic amine of the orphane cytosine, the base left unpaired by extrusion of 8-oxoG from the DNA helix. The cross-link aimed to replace the hydrogen bond between the asparagine and the cytosine. The global structure of hOGG1 in complex with a non-lesion-containing DNA duplex shows overall similarity to that of the 8-oxoG complex. Whereas 8-oxoG fits into the catalytic pocket of the enzyme, normal guanine locates at an "exo-site" 5 Å away from the active site. The extra-helical normal guanine interacts with two active site residues, phenylalanine and histidine, but the contacts are completely different from those of 8-oxoG. This study proposed that hOGG1 is able to read out the subtle structural distinctions between the 8-oxoG versus guanine, allowing the damage admittance to the active site pocket, while rejecting normal bases. Notably, the apo hOGG1 structure suggests that the asparagine position substituted to cysteine for the cross-link is subject to a severe conformational change upon DNA binding [49].

THE H2TH TYPE GLYCOSYLASES: MUTM/FPG, ENDONUCLEASE VIII (ENDOVIII), AND ENDOVIII-LIKE 1 (NEIL1)

These enzymes catalyse the excision of damaged purine bases such as the oxidation products (8-oxoG and 2, 6-diamino-4-hydroxy-5-methylformamidopyrimidine) from dsDNA. Structurally they are H2TH type glycosylases, which include bacterial EndoVIII/Nei, MutM/Fpg, and the mammalian Nei-like proteins, NEIL1-3, and representative structures have been determined for each subgroup [58], [59], [60], [61], [62],[63], [64], [65] and [66]. The helix-multi turn-helix motif was originally discovered in the flap endonuclease FEN-1 structure [16]. The H2TH type glycosylases use a completely different molecular scaffold from the HhH type glycosylases, while the overall topology of these glycosylases is conserved among the family. As seen in the HhH proteins [20], [21], [30] and [35], N- and C-terminal domains create a cleft where DNA is bound, but the H2TH type glycosylases contain α/β structures rather than the primarily α-helical fold of the HhH type glycosylases [59], [60], [61], [62], [63] and [64]. The N-terminal domain has catalytically important amino acids at positions 1–6, followed by two β sheets consisting of four β strands that form an anti-parallel β sandwich flanked by two helices. The C-terminal domain contains the H2TH motif and is helix-rich, typically with the zinc finger contributing the only two β strands. The β-hairpin loop of the zinc finger motif inserts into the DNA minor groove.

MutM, also known as Fpg, is a functional homolog of OGG1 in E. coli and other prokaryotes in that it catalyzes the removal of 8-oxoG from 8-oxoG:C, but MutM binds DNA with a H2TH motif rather than an HhH motif. In contrast to OGG1, MutM has been shown to recognize and remove other oxidized bases [31], [63], [64] and [67]. In structures of the H2TH type glycosylases, which have thus far all been bifunctional N-glycosylases/AP-lyases that recognize oxidized base lesions, an N-terminal proline and glutamate have been identified as catalyzing base excision and β–elimination, while an internal lysine and arginine are proposed to promote an additional activity resulting in a final β,δ-elimination product [59], [60], [61] and [62].

Structures of MutM have been determined with and without DNA, and help elucidate the mechanisms of the H2TH type glycosylases for damage recognition and catalysis [58], [59], [60], [62], [63] and [64]. Structures of an inactive MutM mutant with lesion-containing DNA have shed light on recognition of oxidized bases [63] and [64], particularly in comparison to the hOGG1 complex with DNA carrying 8-oxoG. MutM initiates repair of 8-oxoG, DHU (5, 6-dihydroxyuracil), and a number of other lesions, including FapyG (2,6-diamino-4-hydroxy-5-formamidopyrimidine). As in hOGG1, MutM makes its discriminating contacts with N7 of 8-oxoG, which is contacted by the main chain carbonyl of a serine in MutM. Other protein–based interactions differ from those observed in hOGG1. The structure of MutM with DNA containing DHU revealed that the protonated N3 of DHU overlays with N7 of 8-oxoG and that the DHU carbonyls O2 and O4 are in identical positions to O8 and O6 of 8-oxoG, respectively. In the structure of a complex with FapydG-containing DNA, hydrogen bond contacts are made to the base lesion by four residues, two glutamines, a serine, and an isoleucine, as well as via interactions mediated by structural water molecules [64]. In particular, water-mediated interactions largely discriminate between FapyG and guanine at the five-membered ring of guanine. Only the serine and isoleucine main chain interactions are shared between the recognition mechanisms for FapyG and 8-oxoG. Structures of MutM with reduced abasic DNA and opposite cytosine, guanine, or thymine have shown that a single residue, arginine, specifies the preference for the opposing base by enforcing a hydrogen bonding pattern; cytosine is accommodated in the normal anti position, guanine takes the less favorable syn position, presenting its Hoogstein face, and thymine is pushed away from the side chain of the pseudo-base arginine but is still found in the anti-position [59].

E. coli Endo VIII, also known as Nei, shares significant sequence homology and a similar catalytic mechanism with MutM, though the Endo VIII H2TH glycosylase differs significantly from MutM in substrate specificity (Fig. 2D) [61] and [66]. Endo VIII excises thymine glycol, DHU, -ureidoisobutyric acid, and other oxidized pyrimidines from DNA [68] and [69]. Interestingly, the substrate specificity of Endo VIII overlaps that of HhH type Endo III. Endo VIII is also reported to show some activity towards 8-oxoG, possibly performing a back-up function in the enzymatic system that repairs oxidized purines in DNA. The structure of a covalently trapped Nei–DNA complex has been determined, revealing acid/base chemistry by conserved proline and glutamate in the very N-terminus, consistent with MutM catalysis [61]. To date, however, no structure has been available to show its damage recognition mechanism. By contrast to the complex form, apo Endo VIII is flattened with ~50° rotation of the two major domains pivoting at a hinge domain and indicating a global conformation change from an elongated 'open' form to a 'closed' form upon DNA binding [66].

Human NEIL1 and NEIL2 both have higher affinity for DNA bubble structures, as would be expected to occur during transcription or replication, compared to single or double-stranded DNA [70]. NEIL1 expression is activated during S-phase, when DNA is being synthesized, whereas NEIL2 levels are cell cycle independent, implicating NEIL2 in transcription-coupled repair. The mammalian NEIL proteins are different in their regulation and substrate specificity, though the bacterial and mammalian H2TH proteins share similar overall folds [65] and [66]. The H2TH motifs are used in a manner similar to the HhH, namely to recognize DNA through interactions with the DNA backbone. Interestingly, the structure of the catalytic core of human NEIL1 revealed a "zincless finger" in which the structural motif was three-dimensionally conserved despite missing a true zinc ion binding site (Fig. 2D).

NEIL1 contains a structural motif composed of two anti-parallel -strands that mimics the anti-parallel β-hairpin zinc finger found in other H2TH type glycosylases but lacks the loops to maintain the zinc-binding residues and, therefore, does not coordinate zinc [65]. In contrast, NEIL2 has been shown to bind zinc ion as observed in the other family members [71]. NEIL1 has an extended C-terminal region, which is not included in the crystal structure, which has been implicated in interactions with Pol β and DNA ligase IIIα [72]. Even the processing of the 3' phosphates generated by NEIL1 and NEIL2 differs from other glycosylases of the BER pathway, because it is carried out by polynucleotide kinase rather than AP endonuclease [72].

THE DISTINCT ZINC SITE ROLES IN AGT/ADA PROEINS

Mammalian O^6-alkylguanine-DNA methyltransferase AGT and bacterial Ada are homologous proteins that directly remove alkyl groups at the O6 position of guanine in a stoichiometric irreversible reaction that directly reverses the DNA base damage [73]. As a complement to the BER machinery, the structural biochemistry of these direct damage reversal proteins is also aiding in uncovering unifying themes and specific mechanisms in damaged base recognition. Structures have been determined for bacterial, archaeal, and human AGT/Ada proteins with and without DNA [74], [75], [76], [77], [78] and [79]. Human AGT is of particular interest because it repairs damage induced by some anticancer chemotherapeutics because O^6-alkylguanine is one of the important damaged products that promotes apoptosis and hence therapeutic responses. The crystal structure of apo human AGT, as well as structures of the methylated and benzylated enzymes, revealed a two domain α/β fold (Fig. 2E) [76] and [77]. The N-terminal domain consists of an anti-parallel β sheet followed by two α helices. The C-terminal domain is comprised of a β hairpin, four α helices, and a 3_{10} helix, which harbors a conserved Pro-Cys-His-Arg motif. The C-terminal domain also contains a helix-turn-helix (HTH) motif, often used by DNA binding proteins for sequence-specific recognition [80]. The methylated and benzylated AGT structures, in which a catalytic cysteine has a covalently attached respective alkyl groups, established the active site as being near the recognition helix of the HTH motif.

Interestingly, the DNA-bound structures revealed that the AGT HTH motif is not used as seen in all other existing examples of DNA binding by HTH proteins that bind the recognition helix in the major groove, where it can take a broad range of possible orientations that allow sequence-specific binding [78] and [79]. Instead, the AGT HTH recognition helix lies in the minor groove, which is likely to be advantageous for sequence-independent binding and nucleotide flipping. Another unexpected finding was that binding of AGT to DNA is cooperative and displays directionality, likely to be useful for targeting to areas of localized alkylated damage [78]. Furthermore, analysis of these AGT:DNA structures discovered a 3' phosphate twist mechanism by which a tyrosine is thought to facilitate nucleotide flipping that was also then recognized to be present in other base-flipping systems, including UDG, AlkA, Endonuclease IV, and AP endonuclease [20], [25], [81] and [82].

Human AGT also contains a novel zinc-binding site which is likely to play a structural role [76] and [83]. This zinc atom lies on the active site face with two histidines and two cysteines coordination, but ~20 Å from the reactive cysteine, bridging three strands of the sheet and the coil immediately preceding the domain-spanning helix. The zinc-binding sites are well conserved in higher eukaryotes but not prokaryotes and a leucine, adjacent to one of the histidines, is a site of genetic polymorphism. Apo-metal ion site results in local distortions in the N-terminal domain, including a slight opening of the zinc-binding sites and the vicinity.

Unlike human AGT, E. coli Ada is a unique bifunctional DNA repair protein with an extended N-terminal domain (Fig. 1 and Fig. 2) [84]. The C-terminal domain is homologous to human AGT, transferring alkyl group of O^6-alkylguanine to a cysteine residue, whereas the N-terminal domain repairs the Sp-configured methyl phosphotriester lesion in DNA, also responsible for adaptive response. Structures of human AGT and bacterial Ada C-terminal domain are well conserved despite low primary amino acid sequence homology (Fig. 1) [74]. A loop of AGT, correspond to the first α helix of Ada C-terminal domain, a major difference between the human and bacterial structures. Other differences include the tilting of a helix by ~30°, and the presence of two 3_{10} helices in the AGT structure. The N-terminal domain of Ada is the key component of the chemosensory mechanism by which E. coli monitors its intracellular methylation burden and activates ada regulon [85]. The regulatory regions of ada regulon promoters contain a methylation-dependent activation element comprising conserved AAT and GCAA sequences separated by a six base pair spacer. Methylated N-Ada, resulting from methyl phosphotriester repair stably binds this site and recruits RNA polymerase for the initiation of transcription [85]. The N-terminal domain of Ada contains two domains connected by a flexible linker. Initial domain folds into a globular domain having a central three-strand β sheet sandwiched between a helices, with the zinc ion being bound to four cysteins residues, including a methyl acceptor, at one edge of the sheet, the other domain folds into α–helical bundles similar to that found in the Rob and MarA transcription factors [85]. In structures of DNA complexes (solution and crystal), the elongated N-terminal Ada consisting of two domains lie along one face of the DNA helix. The globular domain is bound over the minor groove, while the bundle domain inserts the recognition helix of its canonical HTH motif into the major groove. The complex structure also shows that both the DNA repair and methylation-dependent DNA binding functions of Ada are dependent on electrostatic switching by a coordinated zinc atom at the four cysteines.

THE BACTERIAL ALKB AND HUMAN HOMOLOGUES HAVE A DISTINCT INTERFACE FOR NUCLEOTIDE BINDING

AlkB and the homologues belong to a superfamily of 2-oxoglutarate- and iron-dependent oxyganases [86]. E. coli AlkB and the human homologues, ABH2, and ABH3 standalone restoring normal bases from alkylation damage in both RNA and DNA [87], [88], [89] and [90]. These oxygenases share some mechanistic features with cytchrome P450 enzymes, which are involved in the process of detoxifying foreign substances. These enzymes require ferrous iron and catalytic hydroxylation reactions where O_2 is the oxygen donor. When N-alkyl group is hydroxylated by the catalytic activity, the resulting alcohol is spontaneously converted to amine and aldehyde. The Fe (II)/2-oxoglutarate-dependent oxygenases couple the substrate oxidation to conversion of 2-oxoglutarate into succinate and CO_2. In addition to 1-alkyladenine and 3-alkylcytosine, which are major cytotoxic lesions, bacterial AlkB is capable of repairing 1-methylguanine, 3-methylthymine, 1, N^6-ethenoadenine, 3, N^4-ethenocytosine and the analogues [91], [92] and [93]. Compared with the bacterial enzyme, the substrate specificity of human homologues is more restricted. Besides 1-methyladenine and 3-methylcytosine, ABH3 shows a much lower activity toward 1, N^6-ethenoadenine. Interestingly, bacterial AlkB and human ABH3 both tend to prefer single-stranded DNA and RNA, whereas human ABH2 is more active on double-stranded substrates [94], [95] and [96].

AlkB crystal structures have revealed the catalytic center with mechanistic implications [97]. Whereas the core domain consisting of double-layered sheets form a "jellyroll" fold closely matches that observed in other deoxygenase superfamily members, AlkB has a unique extra domain made of three strands, designed as nucleotide-recognition lid (Fig. 2F). The cofactors, 2-oxoglutarate and iron, are located at the conserved H-X-D/E-X_n-H motif [98] in the core domain. In a complex structure with trinucleotide carrying 2'-deoxy-1-methyladenosine, the methylated base moiety is stuck in a deep, predominantly hydrophobic cavity, where the nucleoside was pinched out between bulky residues typtophan from the nucleoti de-recognition lid and histidine, an iron-anchoring residue in the oxyganase core. The methyl group directly contacts its molecular surface as well as an oxygen atom on the 1-carboxylate of 2-oxoglutrate. To efficiently repair bulkier alkyl modifications or other modified bases, however, the enzyme is likely to require reorientation of the base and movement of the attached polynucleotide backbone to maintain the precise alignment of the target alkyl atom, suggesting that the substrate-binding lid is flexible to adjust the substrate binding.

At almost the same time, the structure of human homologue, ABH3, was also reported in a complex with iron and 2-oxoglutarate, showing overall structural conservation, especially in the jellyroll core, and nearly identical geometry of iron and 2-oxoglutarate [99] (Fig. 2F). However, significant differences between bacterial and human enzymes occur in the substrate-binding cavity and in other key regions in the vicinity of the active sites. In ABH3, the substrate-binding cavity has more polarity and the key tryptophan of AlkB is substituted to a tyrosine (Fig. 2F), presumably restricting the substrate specificity of the human enzyme. Moreover, the proposed interfaces for nucleotide backbone interactions are poorly conserved and structurally different. Structural diversity is also remarkable at the substrate-binding lid. This region of ABH3 makes unique turns and forms different secondary structure elements. These striking variations in contact residues support distinct binding modes among the AlkB homologues.

In humans, seven homologues of AlkB have been identified but demethylase activity has been observed for only ABH2 and ABH3. The function for the other five homologues remains unclear. Even ABH2 is drastically different from ABH3 in the substrate specificity. Complex structures will shed more light on functional similarities and diversities of the redundant genes.

THE STRUCTURE-SPECIFIC NUCLEASE FEN-1 IN DNA AND PCNA INTERPLAY

The BER pathway, involving nick formation, DNA synthesis and ligation, is dependent not only upon the nature of the glycosylase and its resultant product, but also upon the timing in the cell cycle and the subnuclear localization of the process. Thus, the long-patch pathway shares a set of factors that are also involved in DNA replication, such as DNA pol δ and ε, FEN-1, PCNA, and DNA ligase I, whereas the short-patch repair pathway uses different enzymes including pol β and ligase III stimulated by XRCC1 [7] and [100].

In the BER machinery, modified AP sites are processed via the PCNA-dependent pathway, which involves cleavage of a two to eight nucleotide "flap". FEN-1 binds the ring-shaped adaptor protein, PCNA, to precisely remove flap structures during DNA replication and long-patch BER [16], [101], [102], [103], [104] and [105]. In these contexts, FEN-1 binding to PCNA stimulates FEN-1 endonuclease activity [101]. Structures of FEN-1 show a saddle-shaped, single-domain α/β protein with a deep groove along one face formed from the central seven-stranded β sheet, an anti-parallel β ribbon, and two a-helical bundles (Fig. 3)[16], [102], [103] and [104]. FEN-1 has the conserved PCNA binding domain located near the C-terminus of the enzyme.

The structure of Achaeoglobus fulgidus FEN-1–DNA complex, along with that of the PCNA–FEN-1 peptide complex, reveals the interplay among these three key components (Fig. 3A) [103]. FEN-1 recognizes the flap junction by binding a single, unpaired, 3' nucleotide (3' flap), in a specific pocket ~25 Å from the nuclease active site. Recognition of the 3' flap and associated duplex DNA by two α-loop-α motifs promotes DNA kinking by ~90° and conformational closing of the 5' flap binding motif. These interactions allow FEN-1 precisely position and cleave the 5' flap.

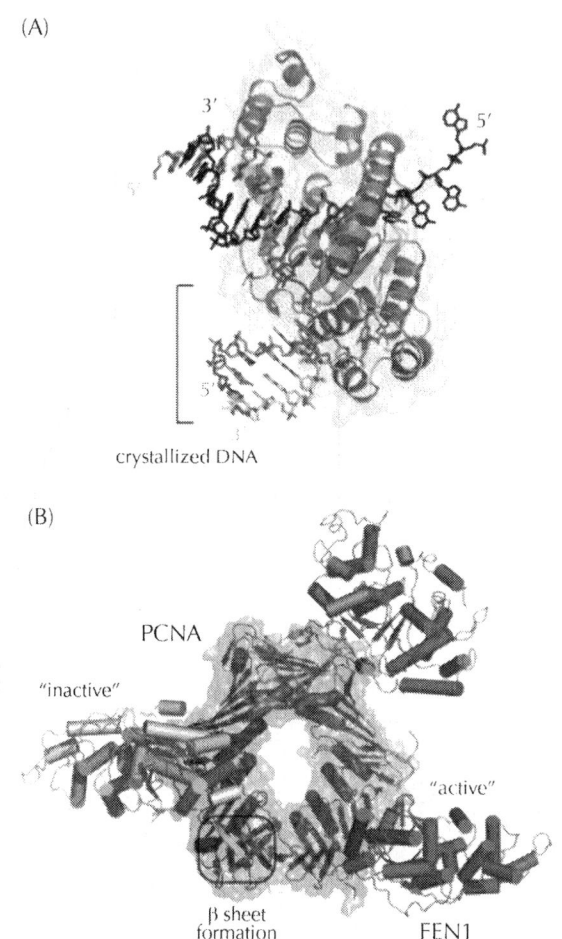

Figure 3: Structures of FEN-1 complexes. FEN-1 in complex with (A) DNA and (B) PCNA.

A crystal structure of a peptide derived from afFEN-1 bound to afPCNA shows that the unstructured C-terminal regions in FEN-1 and PCNA create an intermolecular anti-parallel β sheet interface in the PCNA–FEN-1 peptide complex [103]. The interface directly links adjacent PCNA and DNA binding regions of FEN-1, suggesting a potential mechanism for how PCNA binding affects FEN-1 activity. The crystal structure of human FEN-1 bound to PCNA also shows that the C-terminal domain of FEN-1, which consists of two β-strands connected by a short helix, is a key for the interaction with PCNA, consistent with archaeal PCNA–FEN-1–peptide complex (Fig. 3B) [103] and [104]. Additionally, the full-length enzyme complex reveals another interface in the core domain of FEN-1. These interactions orient the enzyme into an inactive 'locked-down' form, suggesting that a significant rearrangement must occur in a pivotal hinge region to convert from an inactive to an active conformation. In general, the ability of different BER pathway enzymes to bind to PCNA by similar motifs and coupled β-strand interactions prompted the proposal that interface mimicry and exchange allows an ordered sequence of handoffs induced by the specific DNA products, such that the structures of the PCNA:enzyme:DNA complexes themselves help to coordinate individual repair steps into a pathway by promoting specific sequential exchanges without the release of toxic intermediates [103].

THE REPAIR POLYMERASE β AND OPEN/CLOSE SWITCHING FOR BASE FIDELITY

DNA polymerase β of the X family is responsible for DNA synthesis during BER [7] and [106], and in fact, pol β contributes two activities to the BER pathway. A deoxyribose phosphate lyase activity, associated with an amino-terminal 8-kDa lyase domain, excises a deoxyribose phosphate intermediate during repair of abasic sites and generates a 5-phosphate in a single-nucleotide gap. The nucleotidyl transferase activity of pol β is associated with a 31-kDa polymerase domain that fills the single-nucleotide gap and is also needed for some alternate repair pathways that require longer gap-filling DNA synthesis (e.g. long-patch base excision repair) [107], [108] and [109]. The single or multiple nucleotide gap-filling synthesis requires that the polymerase efficiently incorporate correct nucleosides that preserve Watson-Crick base-pairing to maintain genetic integrity since pol β lacks proofreading activity.

Crystallographic structures of pol β have been determined in complex with nicked, gapped, non-gapped, and blunt-end DNA [107], [108], [109], [110], [111] and [112]. Similar to structures of various DNA polymerases derived from diverse sources, the pol β structure is also composed of three functionally distinct domains, C (catalytic), D (duplex DNA binding), and N subdomains (dNTP and template nucleotide binding), referred to as palm, thumb, and fingers subdomains, respectively, according to the nomenclature that utilizes an architectural analogy to a right hand (Fig. 4A). The complex structures show that pol β kinks both gapped and nicked DNA substrates, facilitating access to the 3' end to carry out the gap-filling process. Recent structures of pol β, lacking a proofreading exonuclease activity, in complex with damaged and mispaired DNA reveal its accuracy.

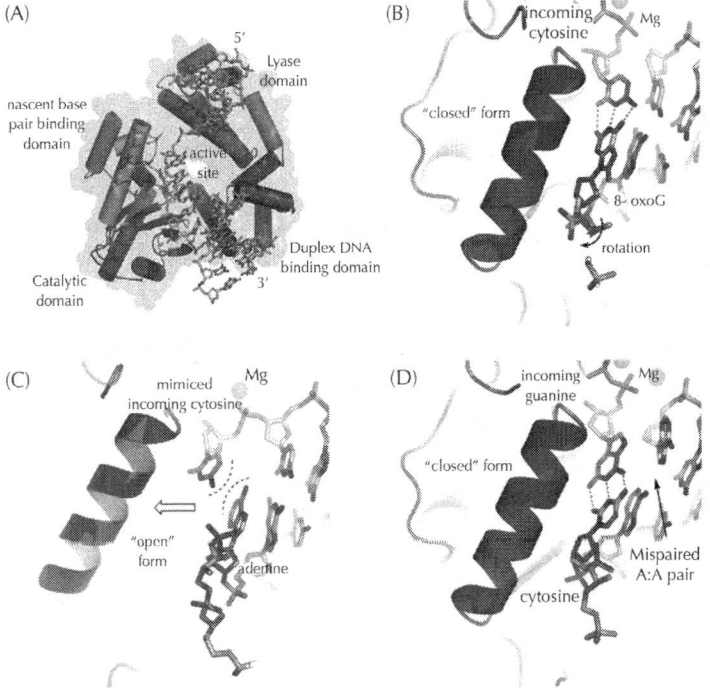

Figure 4: Snap shots of DNA polymerase β give insight into its DNA mismatch repair mechanism. (A) Overall view of pol β in complex with its DNA substrate. (B) Pol β recognizes the oxidized base 8-oxoG. (C) Stalls the incoming mismatch base in the open form of the enzyme. (D) Allows for further DNA extension.

Structure of pol β with DNA carrying 8-oxoG shows that the modified guanine residue is in an anticonformation and forms Watson-Crick hydrogen bonds with an incoming dCTP (Fig. 4B) [113]. In the conformation, the N subdomain of pol β is closed around the nascent base pair and interacts with the N-terminal lyase domain. To accommodate the carbonyl group at C8, however, subtle conformational changes occur both in the deoxyribose and 5'-phosphodiester linkage of the modified nucleotide. The 5'-phosphate backbone of the template nucleotide flips 180° away from the oxygen at C8 of 8-oxoG. This circumvents an unfavorable clash between the base moiety and backbone, thus providing a stable binding of cytosine with anti-8-oxoG in a template. By contrast to dCTP, an incoming dATP forms a staggered conformation with anti-8-oxoG. In this conformation, the N subdomain is in a fully open conformation. This has important implications for pol β function, since the ability to replicate through 8-oxoG lesions depends on maintaining the 8-oxoG base in the anti-conformation.

DNA polymerases distinguish incorrect nucleotides by binding more weakly and inserting more slowly than they do with the correct complementary nucleotide. DNA polymerase-imposed constraints during nucleotide discrimination have remained poorly understood, in part due to an inability to trap an incorrect dNTP that has a much lower binding affinity than the complementary nucleoside. Since pol β retains affinity for a nicked DNA product, it provides an important opportunity to test DNA with a mismatched pair. By placing the mismatched pair at the nick site it was possible to ensure full occupancy of the mismatched nucleoside in the incoming dNTP binding site [114]. The resulting crystal structures of pol β in complex with two different mispairs (A:C and T:C) reveal that the two bases stack partially, rather than pair through nonstandard hydrogen bonds, resulting in the disturbance of the magnesium ion positions (Fig. 4C). This shift in catalytic magnesium ion locations prevents the open/close conformational switching of the N subdomain that is believed to be required for catalytic cycling. The N subdomain of pol β adopts a partially open conformation that is suboptimal for catalysis in the mispaired complexes. This partially open conformation results in distinct hydrogen bonding networks unique for each mispair. In contrast, the structure of an A family DNA polymerase complex with DNA mismatches shows that mismatches are situated at the boundary of the post-insertion site and in many cases, engage hydrogen bonds similar to those observed in DNA duplex alone [115]. In three cases, however, the primer terminus remained in the insertion site. In these situations, an aromatic side chain blocks the template site, and the non-complementary template base is flipped out of the helix axis into the pre-insertion site. Importantly, the structure of pol β complex with mispaired DNA confirms that geometric constraints imposed by the template base occur in part prior to the dNTP binding, whereas these constraints are presumably imposed subsequent to nucleotide binding for A-family DNA polymerase.

Pol β shows relatively higher accuracy compared with low fidelity polymerases, but even pol β is subjected to miscoding. The structure of DNA pol β with a DNA mismatch at the boundary of the polymerase active site reveals how pol β manages a mismatch (Fig. 4D) [116]. The structure of this complex indicates that the template adenine of the mispair stacks with the primer terminus adenine while the coding cytosine is flipped out of the DNA helix. By soaking with dGTP, the crystals of the binary complex resulted in crystals of a ternary substrate complex. In this case, the template cytosine forms Watson-Crick hydrogen bonds with the incoming dGTP in the DNA duplex. The adenine at the primer terminus has rotated into a syn-conformation to interact with the opposite adenine in a planar configuration, yet the 3'-hydroxyl on the primer terminus is out of position for efficient nucleotide insertion.

Pol β shows lyase activity toward the 5'-deoxyribose phosphate group, which occurs in some BER sub-pathways [117]. Pol β was crystallized with 5'-phosphorylated abasic sugar analogs in nicked DNA [118]. However, the crystal structure shows that the 5'-deoxyribose phosphate group is bound in a non-catalytic binding site. In the complex structure, the catalytic nucleophile in the 5'-deoxyribose phosphate lyase reaction and all other potential secondary nucleophiles, are too far away to participate in nucleophilic attack on the C1' of the sugar. This suggests that a rotation of 120°, which pivots the 3'-phosphate of the 5'-deoxyribose3'-phosphate, group is required to position the catalytic lysine close to the 5'-deoxyribose C1'.

THE LIGASE RING AND DNA COMPLEX PARADOX

All BER pathways are completed by ligating the 3' primer terminus of the newly inserted nucleotide(s) to the parent DNA strand. This end-joining reaction, which is catalyzed by DNA ligase, is required by all organisms and serves as the ultimate step of not only DNA repair but also DNA replication and recombination processes. In accordance with their role in catalyzing phosphodiester bond formation, genetic inactivation of DNA ligases causes pleiotropic phenotypes including lethality and hypersensitivity to DNA damaging agents [119]. DNA ligases are members of the nucleotidyl transferase superfamily whose members interact with a nucleotide cofactor to form a covalent enzyme-nucleoside monophosphate. In general, DNA ligase catalyses formation of phosphodiester bonds at single-strand breaks between adjacent 3'-hydroxyl and 5'-phosphate in DNA duplex. For the BER machinery, DNA ligase III carries out this essential ligation step. Ligase III has a zinc finger motif in the N-terminus, which consists of three cysteines and one histidine (Fig. 5A and B) [120]. The enzyme recognizes single-strand nicks and other damage features in dsDNA, both through the N-terminal domain containing a single zinc finger and the catalytic core domain. The NMR structure shows that the zinc finger domain of human DNA ligase IIIβ consists of a three-stranded anti-parallel β sheet following two helices and that zinc atom is coordinated between the β sheet and a helix [121]. This structure is also considered to represent zinc fingers of other enzymes that recognize DNA damage, such as poly (adenosine-ribose) polymerase and nick-sensing DNA 3'-phosphoesterase [122].

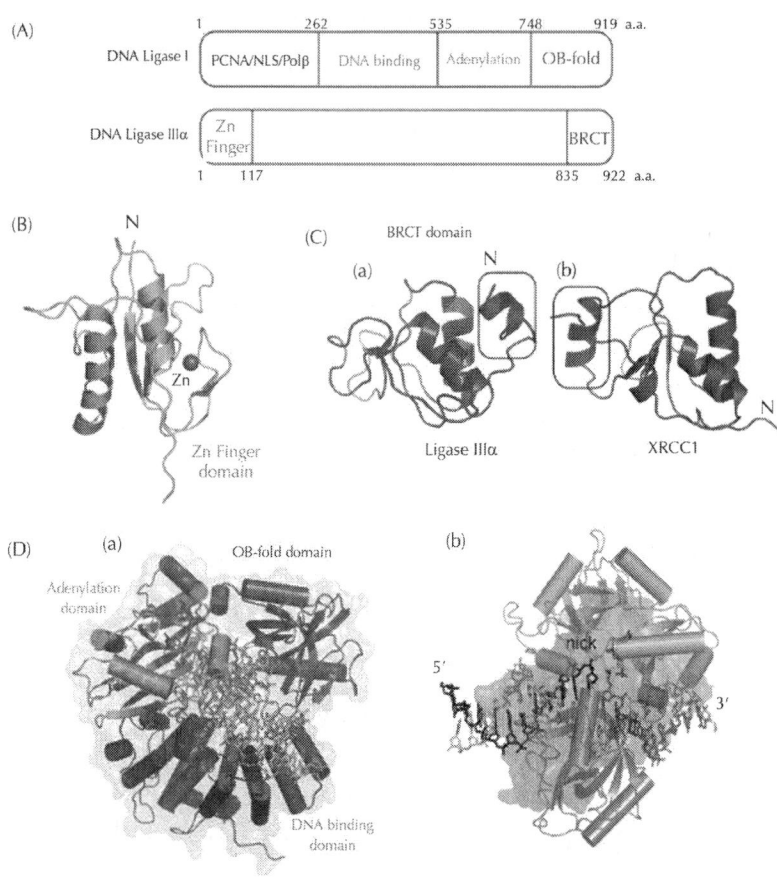

Figure 5: Structures of DNA ligase domains and DNA complex. (A) Domain structures of mammalian DNA ligase I and IIIα. (B) Comparison between ligase IIIα and XRCC1 BRCT domains. (C) Structure of the zinc finger binding domain of ligase IIIα. (D) Structure of ligase I in a complex with DNA.

Ligase III has two isomers, DNA ligase IIIα and β, which differ in domain structure: ligase IIIα consists of an N-terminal DNA-binding domain carrying a zinc finger motif, a core catalytic domain, and a C-teminal BRCT (BRCA1 carboxyl terminus) domain (Fig. 5A), whereas ligase IIIβ lacks the BRCT domain [123]. Widely distributed ligase IIIα is involved in DNA repair in most cells, whereas ligase IIIβ is thought to be involved in completion of homologous recombination during meiotic prophase or post-meiotic DNA repair [124] and [125]. While no structure of the entire DNA ligase III is yet available, domain and homologous structures highlight several of its characteristic features.

The BRCT domain of DNA ligase IIIα interacts with the distal BRCT domain of XRCC1 [126]. It has been suggested that XRCC1 coordinates the rapid recruitment of pol β, poly (ADP-ribose) polymerase, and ligase IIIα to DNA substrate [122]. Ligase IIIα requires XRCC1 for stability, whereas ligase IIIβ, which lacks a BRCT domain is stable without XRCC1. The solution structure of ligase IIIα BRCT domain shows the conserved BRCT fold as also seen in the XRCC1 BRCT domain, which consists of a four-stranded parallel β sheet with a two α helix bundle against one face of the sheet (Fig. 5C) [123]. The major difference is that ligase IIIα is missing ten residues that comprise an additional helix in XRCC1. The truncated DNA ligase IIIα BRCT domain exists as a dimer in solution. It is suggested that the N-termini are involved in forming the homodimer interface, as was similarly observed in the crystallographic structure of the BRCT domain from XRCC1 [123] and [126].

DNA ligase I joins Okazaki fragments during DNA replication but also functions in long-patch BER. The recent crystallographic structure of human DNA ligase I in complex with a nicked, 5' adenylated DNA intermediate reveals that the enzyme redirects the path of the double helix to expose the nick termini for the strand-joining reaction (Fig. 5D). This ligase:DNA complex furthermore shows a DNA-binding domain that stabilizes the DNA in a distorted structure, and positions the catalytic core on the nick [127].

To obtain a stable reaction intermediate complex, a synthetic nicked duplex terminated with a 3' dideoxynucleotide was used, thereby blocking the final sealing step. The structure shows that three domains of ligase I completely encircle DNA. The N-terminal DNA binding domain (DBD) consisting of 12 helices provides most of the interactions with DNA, localized extensively in the minor groove, and also contacts the adenylation domain (AdD) and the OB-fold domain (OBD). The DBD binds DNA evenly on both sides of the nick with a ~2-fold axis of symmetry. The DBD is also conserved in ligase III and IV. The AdD mixed α/β fold has resemblance to the nucleotide binding domains of other known ligase structures with conserved active site sequences. In the complex, the 5' AMP binds deep within a pocket of the AdD, outside of the DNA duplex. Downstream of the nick, DNA is in the B-form, which may explain why ligase I does not seal Okazaki fragments before the RNA primer is removed, whereas upstream DNA adopts A-form with an expanded minor groove. The OBD binds in the minor groove adjacent to the ends of the nicked DNA, and alters the curvature of the DNA. The OBD distorts the DNA, leading to an underwound conformation that widens the major and minor grooves, and causes a more than 5 Å shift in the helical axis between upstream and downstream segments. These distortions bring the nick close to the active site and also explain the fidelity of the DNA ligase.

A conserved motif of the OBD, which is involved in the initial enzyme-AMP formation, is distant from the other essential motifs in the AdD. The coordination of the C-terminal OBD in the ligase I complex with DNA implies that a rearrangement between the OBD and AdD is necessary to proceed from enzyme-AMP to a DNA-bound form. Recent results on a ligase derived from Pyrococcus furiosus offer insightful observations on the initial ATP processing [128]. The archaeal enzyme is also an ATP-dependent DNA ligase, consisting of three major domains: an N-terminal DNA binding domain, an adenylation domain, and a C-terminal OB-fold domain. The crystal structure of the P. furiosus ligase shows a closed conformation of the two domains at the carboxyl terminus, creating a small compartment where the AMP molecule is non-covalently bound. The closed conformation is formed by part of the enzyme mimicking DNA backbone. The identified residues are well conserved between archaeal and human enzymes.

Although mammalian ligases show homology in the primary sequences and are predicted to catalyze ligation in a similar manner, each enzyme has distinct roles in a cell. Understanding the specific interactions with other cellular components, such as ligase I/PCNA or ligase III/XRCC1 complexes, will provide insights into the complexities of ligase structure–function relationships needed to explain its distinct biological roles. Obvious critical issues include how the ligase ring can load onto DNA, the role of PCNA in this loading process, and whether ligase loading and release is general or specific to the individual ligase classes.

THE DIVERSE CHEMICAL ANATOMY OF BER AND GENOME MAINTENANCE

As outlined above the three-dimensional structures of DNA N-glycosylases, N-glycosylase/AP-lyases, structure-specific nucleases, repair polymerase, DNA ligase, and PCNA tethering complexes provide unifying concepts while also revealing mechanistic diversity with important biological implications. As we develop systematic structural characterizations of BER complexes we will gain an informed understanding of the mechanisms by which these BER proteins function to specifically recognize, remove, and repair DNA base damage without the release of toxic and mutagenic intermediates.

Beginning with the discoveries of the HhH sequence-independent DNA binding motif [15] and both nucleotide flipping to allow specific damage recognition [13], [20], [21], [30] and [35] and dinucleotide flipping to expose the opposite base [81], initial BER complex structures have revealed the power of structural biology to help define the structural basis for specificity and catalysis [12] and [29]. These structures have furthermore suggested how dynamic assemblies may act in efficient damage scanning and detection [78], that initial enzyme–substrate strain promotes dissociative mechanisms [25], and that the resulting product inhibition aids handoffs without the release of toxic intermediates [82] and [110]. Particularly surprising and intriguing structures have furthermore discovered the roles of interface mimicry (where interface motifs and components resemble one another sufficiently to be exchanged) in pathway control and interface exchange [129], [130] and [131], and the critical controls from conformational switching [103].

These existing and emerging lesions learned from BER machinery are proving to be prototypes for other DNA repair pathways. We find conformational switching applying to other DNA repair pathways including transcription-coupled nucleotide excision repair [132], Holliday junction resolution [133], and double strand break repair [134] and [135]. Similarly interface mimicry in pathway regulation is showing up for other repair pathways [136]. In any case, the detailed understanding of the molecular basis for BER will continue to be important in its own right and in the understanding of mutations, cancer initiation and cancer predispositions.

As the spontaneous mutation rate in somatic cells cannot account for the multiple mutations observed in human cancers, a compelling model for cancer initiation involves the development of a 'mutator phenotype'. Defects in base excision repair are obviously among the key components that could cause such a mutator phenotype. Indeed, mutations to alter the specificity of UDG were used to create just such a mutator phenotype by design [137]. Consistent with the connection of repair defects with cancer, deficiencies in mismatch repair and nucleotide excision repair can cause cancer predispositions [138] and [139]. Why have not many BER defects been shown to cause cancers so far? We suggest that BER deficiencies are not likely to leave specific fingerprints in terms of DNA defects as they primarily will cause single base changes. This makes the search for a specific BER defect linked to specific cancers much more difficult than was the case for mismatch repair and satellite instabilities, for example. Yet, screening for low activities of particular DNA glycosylases or other BER enzymes may not be the most useful approach, since as evident from the motifs, interfaces, and conformational changes outlined here, we expect many key defects may be in specificity, fidelity, handoffs, and the coordination of different repair steps.

The direct connection of the BER glycosylase MYH to cancer resulted in part from having an easily recognized phenotype (the development of hundreds to thousands of adenomatous polyps throughout the intestinal tract indicative of Familial adenomatous polyposis). However, this discovery has served to highlight the value of obtaining a detailed understanding of BER structural biology including the mechanisms for DNA base damage recognition and removal plus the potential relationship of these processes to the molecular basis for human cancers and other pathologies. In fact, emerging results suggest that DNA repair proteins may be master keys for aging and brain pathology as well as cancer [140]. For these reasons, it will be important to have future BER structures and chemical tools that will test proposals for the detailed basis for damage recognition and removal including substrate assisted catalysis in oxidized base repair [14], [39], [40], [41], [42], [43], [44], [45], [46], [47], [48], [49], [50], [51] and [141]. Multiple structures will also be needed to understand functionally important conformational changes, such as how the ligase ring loads onto PCNA and DNA [96], [142] and [143] and the possible roles of conformational switching in this process. Defining the structural biochemistry of mitochondrial DNA polymerase and its interactions and conformations controlling processivity in repair synthesis is likely important for understanding aging and neurodegenerative diseases associated with DNA repair processes [144]. Interface mimicry and its role in interface exchanges such as seen for PCNA [89] and [145] also merits systematic structural characterizations. Such detailed structures of substrate and product complexes and coordinating assemblies will ultimately provide answers to how proteins are joined to pathways and why the evidently redundant specificities in BER enzymes are important for genome integrity, aging, and cancer avoidance.

ACKNOWLEDGMENTS

The work on DNA Repair in the Authors' laboratory is supported by the Human Frontiers in Science Program (S.I. and J.A.T.), by NIH grants GM46312, and CA92584 (J.A.T) and the Japan Society for the Promotion of Science (K.H.). We thank Dr. Brian R. Chapados for the FEN-1–DNA complex coordinate and critical comments on this manuscript, and Drs. Jill Fuss and Julie L. Tubbs for critical reading of the manuscript.

REFERENCES

1. D.R. Denver, S.L. Swenson, M. Lynch, An evolutionary analysis of the helix-hairpin-helix superfamily of DNA repair glycosylases, Mol. Biol. Evol. 20 (2003) 1603– 1611.
2. H.E. Krokan, R. Standal, G. Slupphaug, DNA glycosylases in the base excision repair of DNA, Biochem. J. 325 (1997) 1–16.
3. T. Lindahl, R.D. Wood, Quality control by DNA repair, Science 286 (1999) 1897–1905.
4. B. Demple, L.B. Harrison, Repair of oxidative damage to DNA: enzymology and biology, Annu. Rev. Biochem. 63 (1994) 915–948.
5. R.D. Wood, DNA repair in eukaryotes, Annu. Rev. Biochem. 65 (1996) 135–167.
6. J.L. Huffman, O. Sundheim, J.A. Tainer, DNA base damage recognition and removal: new twists and grooves, Mutat. Res. 577 (2005) 55–76.
7. T. Izumi, L.R. Wiederhold, G. Roy, R. Roy, A. Jaiswal, K.K. Bhakat, S. Mitra, T.K. Hazra, Mammalian DNA base excision repair proteins: their interactions and role in repair of oxidative DNA damage, Toxicology 193 (2003) 43–65.
8. S. Boiteux, M. Guillet, Abasic sites in DNA: repair and biological consequences in Saccharomyces cerevisiae, DNA Repair (Amst) 3 (2004) 1–12.
9. R.W. Sobol, M. Kartalou, K.H. Almeida, D.F. Joyce, B.P. Engelward, J.K. Horton, R. Prasad, L.D. Samson, S.H. Wilson, Base excision repair intermediates induce p53-independent cytotoxic and genotoxic responses, J. Biol. Chem. 278 (2003) 39951–39959.

10. B. Kavli, G. Slupphaug, C.D. Mol, A.S. Arvai, S.B. Peterson, J.A. Tainer, H.E. Krokan, Excision of cytosine and thymine from DNA by mutants of human uracil-DNA glycosylase, EMBO J. 15 (1996) 3442–3447.
11. T. Lindahl, Instability and decay of the primary structure of DNA, Nature 362 (1993) 709–715.
12. C.D. Mol, A.S. Arvai, G. Slupphaug, B. Kavli, I. Alseth, H.E. Krokan, J.A. Tainer, Crystal structure and mutational analysis of human uracil-DNA glycosylase: structural basis for specificity and catalysis, Cell 80 (1995) 869–878.
13. G. Slupphaug, C.D. Mol, B. Kavli, A.S. Arvai, H.E. Krokan, J.A. Tainer, A nucleotide-flipping mechanism from the structure of human uracil-DNA glycosylase bound to DNA, Nature 384 (1996) 87–92.
14. C.T. McMurray, J.A. Tainer, Cancer, cadmium and genome integrity, Nat. Genet. 34 (2003) 239–241.
15. M.M. Thayer, H. Ahern, D. Xing, R.P. Cunningham, J.A. Tainer, Novel DNA binding motifs in the DNA repair enzyme endonuclease III crystal structure, EMBO J. 14 (1995) 4108–4120.
16. D.J. Hosfield, C.D. Mol, B. Shen, J.A. Tainer, Structure of the DNA repair and replication endonuclease and exonuclease FEN-1: coupling DNA and PCNA binding to FEN-1 activity, Cell 95 (1998) 135–146.
17. C.F. Kuo, D.E. McRee, C.L. Fisher, S.F. O'Handley, R.P. Cunningham, J.A. Tainer, Atomic structure of the DNA repair
18. 4Fe-4S. enzyme endonuclease III, Science 258 (1992) 434–440.
19. Y. Guan, R.C. Manuel, A.S. Arvai, S.S. Parikh, C.D. Mol, J.H. Miller, R.S. Lloyd, J.A. Tainer, MutY catalytic core, mutant and bound adenine structures define specificity for DNA repair enzyme superfamily, Nat. Struct. Biol. 5 (1998) 1058–1064.
20. C.D. Mol, A.S. Arvai, T.J. Begley, R.P. Cunningham, J.A. Tainer, Structure and activity of a thermostable thymine-DNA glycosylase: evidence for base twisting to remove mismatched

normal DNA bases, J. Mol. Biol. 315 (2002) 373–384.

21. T. Hollis, Y. Ichikawa, T. Ellenberger, DNA bending and a flip-out mechanism for base excision by the helix-hairpin-helix DNA glycosylase, Escherichia coli AlkA, EMBO J. 19 (2000) 758–766.

22. S.D. Bruner, D.P. Norman, G.L. Verdine, Structural basis for recognition and repair of the endogenous mutagen 8-oxoguanine in DNA, Nature 403 (2000) 859–866.

23. D.E. Volk, P.G. House, V. Thiviyanathan, B.A. Luxon, S. Zhang, R.S. Lloyd, D.G. Gorenstein, Structural similarities between MutT and the C-terminal domain of MutY, Biochemistry 39 (2000) 7331–7336.

24. B. Hendrich, U. Hardeland, H.-H. Ng, J. Jiricny, A. Bird, The thymine glycosylase MBD4 can bind to the product of deamination at methylated CpG sites, Nature 401 (1999) 301–304.

25. F. Petronzelli, A. Riccio, G.D. Markham, S.H. Seeholzer, M. Genuardi, M. Karbowski, A.T. Yeung, Y. Matsumoto, A. Bellacosa, Investigation of the substrate spectrum of the human mismatch-specific DNA N-glycosylase MED1 (MBD4): fundamental role of the catalytic domain, J. Cell. Physiol. 185 (2000) 473–480.

26. S.S. Parikh, G. Walcher, G.D. Jones, G. Slupphaug, H.E. Krokan, G.M. Blackburn, J.A. Tainer, Uracil-DNA glycosylase-DNA substrate and product structures: conformational strain promotes catalytic efficiency by coupled stereoelectronic effects, Proc. Natl. Acad. Sci. U.S.A. 97 (2000) 5083–5088.

27. I. Ohki, N. Shimotake, N. Fujita, J.-G. Jee, T. Ikegami, M. Nakao, M. Shirakawa, Solution structure of the methyl-CpG binding domain of human MBD1 in complex with methylated DNA, Cell 105 (2001) 487–497.

28. R.I.D. Wakefield, B.O. Smith, X. Nan, A. Free, A. Soteriou, D. Uhrin, A.P. Bird, P.N. Barlow, The solution structure of the domain from MeCP2 that binds to methylated DNA, J. Mol. Biol. 291 (1999) 1055–1065.

29. F. Petronzelli, A. Riccio, G.D. Markham, S.H. Seeholzer, J. Stoerker, M. Genuardi, et al., Biphasic kinetics of the human DNA repair protein MED1 (MBD4), a mismatch-specific DNA N-glycosylase, J. Biol. Chem. 275 (2000) 32422–32429.
30. S.S. Parikh, C.D. Mol, G. Slupphaug, S. Bharati, H.E. Krokan, J.A. Tainer, Base excision repair initiation revealed by crystal structures and binding kinetics of human uracil-DNA glycosylase with DNA, EMBO J. 17 (1998) 5214–5226.
31. J.C. Fromme, A. Banerjee, S.J. Huang, G.L. Verdine, Structural basis for removal of adenine mispaired with 8-oxoguanine by MutY adenine DNA glycosylase, Nature 427 (2004) 652–656.
32. M.L. Michaels, J.H. Miller, The GO system protects organisms from the mutagenic effect of the spontaneous lesion 8-hydroxyguanine (7,8-dihydro-8-oxoguanine), J. Bacteriol. 174 (1992) 6321–6325.
33. A.S. Mildvan, D.J. Weber, C. Abeygunawardana, Solution structure and mechanism of the MutT pyrophosphohydrolase, Adv. Enzymol. Relat. Areas Mol. Biol. 73 (1999) 183–207.
34. A. Gogos, J. Cillo, N.D. Clarke, A.-L. Lu, Specific recognition of A/G and A/7,8-dihydro-8-oxoguanine (8-oxoG) mismatches by Escherichia coli MutY: removal of the C-terminal domain preferentially affects A/8-oxoG recognition, Biochemistry 35 (1996) 16665–16671.
35. D.M. Noll, A. Gogos, J.A. Granek, N.D. Clarke, The C-terminal domain of the adenine-DNA glycosylase MutY confers specificity for 8-oxoguanine.adenine mispairs and may have evolved from MutT, an 8-oxo-dGTPase, Biochemistry 38 (1999) 6374–6379.
36. J.C. Fromme, G.L. Verdine, Structure of a trapped endonuclease III-DNA covalent intermediate, EMBO J. 22 (2003) 3461–3471.
37. M.A. Massiah, V. Saraswat, H.F. Azurmendi, A.S. Mildvan, Solution structure and NH exchange studies of the MutT pyrophosphohydrolase complexed with $Mg(2+)$ and 8-oxo-

dGMP, a tightly bound product, Biochemistry 42 (2003) 10140–10154.

38. N. Al-Tassan, N.H. Chmiel, J. Maynard, N. Fleming, A.L. Livingston, G.T. Williams, A.K. Hodges, D.R. Davies, S.S. David, J.R. Sampson, J.P. Cheadle, Inherited variants of MYH associated with somatic G:C→T:A mutations in colorectal tumors, Nat. Genet. 30 (2002) 227–232.

39. N.H. Chmiel, A.L. Livingston, S.S. David, Insight into the functional consequences of inherited variants of the hMYH adenine glycosylase associated with colorectal cancer: complementation assays with hMYH variants and pre-steady-state kinetics of the corresponding mutated E. coli enzymes, J. Mol. Biol. 327 (2003) 431– 443.

40. T.C. Brown, J. Jiricny, A specific mismatch repair event protects mammalian cells from loss of 5-methylcytosine, Cell 50 (1987) 945–950.

41. U. Hardeland, M. Bentele, T. Lettieri, R. Steinacher, J. Jiricny, P. Schar, Thymine DNA glycosylase, Prog. Nucleic Acid Res. Mol. Biol. 68 (2001) 235–253.

42. P. Gallinari, J. Jiricny, A new class of uracil-DNA glycosylases related to human thymine-DNA glycosylase, Nature 383 (1996) 735–738.

43. T.E. Barrett, R. Savva, G. Panayotou, T. Barlow, T. Brown, J. Jiricny, L.H. Pearl, Crystal structure of a G:T/U mismatch-specific DNA glycosylase: mismatch recognition by complementary-strand interactions, Cell 92 (1998) 117–129.

44. L.H. Pearl, Structure and function in the uracil-DNA glycosylase superfamily, Mutat. Res. 460 (2000) 165–181.

45. T.E. Barrett, O.D. Scharer, R. Savva, T. Brown, J. Jiricny, G.L. Verdine, L.H. Pearl, Crystal structure of a thwarted mismatch glycosylase DNA repair complex, EMBO J. 18 (1999) 6599–6609.

46. U. Hardeland, R. Steinacher, J. Jiricny, P. Schar, Modification of the human thymine-DNA glycosylase by ubiquitin-like

proteins facilitates enzymatic turnover, EMBO J. 21 (2002) 1456–1464.

47. R. Steinacher, P. Schar, Functionality of human thymine DNA glycosylase requires SUMO-regulated changes in protein conformation, Curr. Biol. 15 (2005) 616–623.

48. D. Baba, N. Maita, J.G. Jee, Y. Uchimura, H. Saitoh, K. Sugasawa, F. Hanaoka, H. Tochio, H. Hiroaki, M. Shirakawa, Crystal structure of thymine DNA glycosylase conjugated to SUMO-1, Nature 435 (2005) 979–982.

49. J. Song, L.K. Durri, T.A. Wilkinson, T.G. Krontiris, Y. Chen, Identification of a SUMO-binding motif that recognizes SUMO-modified proteins, Proc. Natl. Acad. Sci. U.S.A. 101 (2004) 14373–14378.

50. M. Bjøras, E. Seeberg, L. Luna, L.H. Pearl, T.E. Barrett, Reciprocal "flipping" underlies substrate recognition and catalytic activation by the human 8-oxo-guanine DNA glycosylase, J. Mol. Biol. 317 (2002) 171–177.

51. D.P. Norman, S.J. Chung, G.L. Verdine, Structural and biochemical exploration of a critical amino acid in human 8-oxoguanine glycosylase, Biochemistry 42 (2003) 1564–1572.

52. J.C. Fromme, S.D. Bruner, W. Yang, M. Karplus, G.L. Verdine, Product-assisted catalysis in base-excision DNA repair, Nat. Struct. Biol. 10 (2003) 204–211.

53. S.J. Chung, G.L. Verdine, Structures of end products resulting from lesion processing by a DNA glycosylase/lyase, Chem. Biol. 11 (2004) 1643–1649.

54. A. Banerjee, W. Yang, M. Karplus, G.L. Verdine, Structure of a repair enzyme interrogating undamaged DNA elucidates recognition of damaged DNA, Nature 434 (2005) 612–618.

55. J. Tchou, H. Kasai, S. Shibutani, M.H. Chung, J. Laval, A.P. Grollman, S. Nishimura, 8-Oxoguanine (8-hydroxyguanine) DNA glycosylase and its substrate specificity, Proc. Natl. Acad. Sci. U.S.A. 88 (1991) 4690–4694.

56. J. Park, L. Chen, M.S. Tockman, A. Elahi, P. Lazarus, The human 8-oxoguanine DNA N-glycosylase 1 (hOGG1) DNA repair enzyme and its association with lung cancer risk, Pharmacogenetics 14 (2004) 103–109.
57. L. Le Marchand, T. Donlon, A. Lum-Jones, A. Seifried, L.R. Wilkens, Association of the hOGG1 Ser326Cys polymorphism with lung cancer risk, Cancer Epidemiol. Biomarkers Prev. 11 (2002) 409–412.
58. I. Morland, L. Luna, E. Gustad, E. Seeberg, M. Bjoras, Product inhibition and magnesium modulate the dual reaction mode of hOgg1, DNA Repair (Amst) 4 (2005) 381–387.
59. M. Sugahara, T. Mikawa, T. Kumasaka, M. Yamamoto, R. Kato, K. Fukuyama, Y. Inoue, S. Kuramitsu, Crystal structure of a repair enzyme of oxidatively damaged DNA, MutM (Fpg), from an extreme thermophile, Thermus thermophilus HB8, EMBO J. 19 (2000) 3857–3869.
60. J.C. Fromme, G.L. Verdine, Structural insights into lesion recognition and repair by the bacterial 8-oxoguanine DNA glycosylase MutM, Nat. Struct. Biol. 9 (2002) 544–552.
61. R. Gilboa, D.O. Zharkov, G. Golan, A.S. Fernandes, S.E. Gerchman, E. Matz, J.H. Kycia, A.P. Grollman, G. Shoham, Structure of formamidopyrimidine-DNA glycosylase covalently complexed to DNA, J. Biol. Chem. 277 (2002) 19811–19816.
62. D.O. Zharkov, G. Golan, R. Gilboa, A.S. Fernandes, S.E. Gerchman, J.H. Kycia, R.A. Rieger, A.P. Grollman, G. Shoham, Structural analysis of an Escherichia coli endonuclease VIII covalent reaction intermediate, EMBO J. 21 (2002) 789–800.
63. L. Serre, K. Pereira de Jesus, S. Boiteux, C. Zelwer, B. Castaing, Crystal structure of the Lactococcus lactis formamidopyrimidine-DNA glycosylase bound to an abasic site analogue-containing DNA, EMBO J. 21 (2002) 2854–2865.
64. J.C. Fromme, G.L. Verdine, DNA lesion recognition by the bacterial repair enzyme MutM, J. Biol. Chem. 278 (2003)

51543–51548.

65. F. Coste, M. Ober, T. Carell, S. Boiteux, C. Zelwer, B. Castaing, Structural basis for the recognition of the FapydG lesion (2,6-diamino-4-hydroxy-5-formamidopyrimidine) by formamidopyrimidine-DNA glycosylase, J. Biol. Chem. 279 (2004) 44074–44083.

66. S. Doublie, V. Bandaru, J.P. Bond, S.S. Wallace, and The crystal structure of human endonuclease VIII-like 1 (NEIL1) reveals a zincless finger motif required for glycosylase activity, Proc. Natl. Acad. Sci. U.S.A. 101 (2004) 10284–10289.

67. G. Golan, D.O. Zharkov, H. Feinberg, A.S. Fernandes, E.I. Zaika, J.H. Kycia, A.P. Grollman, G. Shoham, Structure of the uncomplexed DNA repair enzyme endonuclease VIII indicates significant interdomain flexibility, Nucleic Acids Res. 33 (2005) 5006–5016.

68. S. Boiteux, E. Gajewski, J. Laval, M. Dizdaroglu, Substrate specificity of the Escherichia coli Fpg protein (formamidopyrimidine-DNA glycosylase): excision of purine lesions in DNA produced by ionizing radiation or photosensitization, Biochemistry 31 (1992) 106–110.

69. H. Miller, A.S. Fernandes, E. Zaika, M.M. McTigue, M.C. Torres, M. Wente, C.R. Iden, A.P. Grollman, Stereoselective excision of thymine glycol from oxidatively damaged DNA, Nucleic Acids Res. 32 (2004) 338–345.

70. A. Katafuchi, T. Nakano, A. Masaoka, H. Terato, S. Iwai, F. Hanaoka, H. Ide, Differential specificity of human and Escherichia coli endonuclease III and VIII homologues for oxidative base lesions, J. Biol. Chem. 279 (2004) 14464–14471.

71. H. Dou, S. Mitra, T.K. Hazra, Repair of oxidized bases in DNA bubble structures by human DNA glycosylases NEIL1 and NEIL2, J. Biol. Chem. 278 (2003) 49679–49684.

72. A. Das, L. Rajagopalan, V.S. Mathura, S.J. Rigby, S. Mitra, T.K. Hazra, Identification of a zinc finger domain in the human

NEIL2 (Nei-like-2) protein, J. Biol. Chem. 279 (2004) 47132–47138.
73. L. Wiederhold, J.B. Leppard, P. Kedar, F. Karimi-Busheri, A. Rasouli-Nia, M. Weinfeld, A.E. Tomkinson, T. Izumi, R. Prasad, S.H. Wilson, S. Mitra, T.K. Hazra, AP endonuclease-independent DNA base excision repair in human cells, Mol. Cell 15 (2004) 209–220.
74. A.E. Pegg, Repair of O (6)-alkylguanine by alkyltransferases, Mutat. Res. 462 (2000) 83–100.
75. M.H. Moore, J.M. Gulbis, E.J. Dodson, B. Demple, P.C. Moody, Crystal structure of a suicidal DNA repair protein: the Ada O6-methylguanine-DNA methyltransferase from E. coli, EMBO J. 13 (1994) 1495–1501.
76. H. Hashimoto, T. Inoue, M. Nishioka, S. Fujiwara, M. Takagi, T. Imanaka, Y. Kai, Hyperthermostable protein structure maintained by intra and inter-helix ion-pairs in archaeal O6-methylguanine-DNA methyltransferase, J. Mol. Biol. 292 (1999) 707–716.
77. D.S. Daniels, C.D. Mol, A.S. Arvai, S. Kanugula, A.E. Pegg, J.A. Tainer, Active and alkylated human AGT structures: a novel zinc site, inhibitor and extrahelical base binding, EMBO J. 19 (2000) 1719–1730.
78. J.E. Wibley, A.E. Pegg, P.C. Moody, Crystal structure of the human O(6)-alkylguanine-DNA alkyltransferase, Nucleic Acids Res. 28 (2000) 393–401.
79. D.S. Daniels, T.T. Woo, K.X. Luu, D.M. Noll, N.D. Clarke, A.E. Pegg, J.A. Tainer, DNA binding and nucleotide flipping by the human DNA repair protein AGT, Nat. Struct. Mol. Biol. 11 (2004) 714–720.
80. E.M. Duguid, P.A. Rice, C. He, the structure of the human AGT protein bound to DNA and its implications for damage detection, J. Mol. Biol. 350 (2005) 657–666.
81. R. Wintjens, M. Rooman, Structural classification of HTH DNA-binding domains and protein–DNA interaction modes, J. Mol. Biol. 262 (1996) 294–313.

82. D.J. Hosfield, Y. Guan, B.J. Haas, R.P. Cunningham, J.A. Tainer, Structure of the DNA repair enzyme endonuclease IV and its DNA complex: double-nucleotide flipping at abasic sites and three-metal-ion catalysis, Cell 98 (1999) 397–408.
83. C.D. Mol, T. Izumi, S. Mitra, J.A. Tainer, DNA-bound structures and mutants reveal abasic DNA binding by APE1 and DNA repair coordination, Nature 403 (2000) 451–456.
84. J.J. Rasimas, S. Kanugula, P.M. Dalessio, I.J. Ropson, M.G. Fried, A.E. Pegg, Effects of zinc occupancy on human O6-alkylguanine-DNA alkyltransferase, Biochemistry 42 (2003) 980–990.
85. L.C. Myers, M.P. Terranova, A.E. Ferentz, G. Wagner, G.L. Verdine, Repair of DNA methylphosphotriesters through a metalloactivated cysteine nucleophile, Science 261 (1993) 1164–1167.
86. C. He, J. Hus, L. Sun, P. Zhou, D. Norman, V. Dotsch, H. Wei, ̈J. Gross, W. Lane, G. Wagner, A methylation-dependent electrostatic switch controls DNA repair and transcriptional activation by Ada, Mol. Cell 20 (2005) 117–129.
87. L. Arvind, E.V. Koonin, The DNA repair protein AlkB, EGL-9, and leprecan define new families of 2-oxoglutarate- and iron-dependent dioxygenases, Genome Biol. 2 (2001) 0007.1–0007.8.
88. S.C. Trewick, T.F. Henshaw, R.P. Hausinger, T. Lidahl, B. Sedgwick, Oxidative demethylation by Escherichia coli AlkB directly reverts DNA base damage, Nature 419 (2002) 174–178.
89. P.Ø. Falnes, R.F. Johansen, E. Seeberg, AlkB-mediated oxidative demethylation reverses DNA damage in Escherichia coli, Nature 419 (2002) 178–182.
90. T. Duncan, S.C. Trewick, P. Koivisto, P.A. Bates, T. Lindahl, B. Sedgwick, Reversal of DNA alkylation damage by two human dioxygenases, Proc. Natl. Acad. Sci. U.S.A. 99 (2002) 16660–16665.

91. P.A. Aas, M. Otterlei, P.Ø. Falnes, C.B. Vagbø, F. Skorpen, M. ˚Akbari, O. Sundheim, M. Bjøras, G. Slupphaug, E. Seeberg, ˚H.E. Krokan, Human and bacterial oxidative demethylases repair alkylation damage in both RNA and DNA, Nature 421 (2003) 859–863.
92. P.Ø. Falnes, Repair of 3-methylthymine and 1-methylguanine lesions by bacterial and human AlkB proteins, Nucleic Acids Res. 32 (2004) 6260–6267.
93. Y. Mishina, C.G. Yang, C. He, Direct repair of the exocyclic DNA adduct 1, N6-ethenoadenine by the DNA repair AlkB proteins, J. Am. Chem. Soc. 127 (2005) 14594–14595.
94. J.C. Delaney, L. Smeester, C. Wong, L.E. Frick, K. Taghizadeh, J.S. Wishnok, C.L. Drennan, L.D. Samson, J.M. Essigmann, AlkB reverses etheno DNA lesions caused by lipid oxidation in vitro and in vivo, Nat. Struct. Mol. Biol. 12 (2005) 855–860.
95. Y. Mishina, C. He, Probing the structure and function of the Escherichia coli DNA alkylation repair AlkB protein through chemical cross-linking, J. Am. Chem. Soc. 125 (2003) 8730–8731.
96. P.Ø. Falnes, M. Bjøras, P.A. Aas, O. Sundheim, E. Seeberg, ˚ Substrate specificities of bacterial and human AlkB proteins, Nucleic Acids Res. 32 (2004) 3456–3461.
97. Y. Mishina, C.-H.J. Lee, C. He, Interaction of human and bacterial AlkB proteins with DNA as probed through chemical cross-linking studies, Nucleic Acids Res. 32 (2004) 1548–1554.
98. B. Yu, W.C. Edstrom, J. Benach, Y. Hamuro, P.C. Weber, B.R. Gibney, J.F. Hunt, Crystal structures of catalytic complexes of the oxidative DNA/RNA repair enzyme AlkB, Nature 439 (2006) 879–884.
99. R.P. Hausinger, FeII/alpha-ketoglutarate-dependent hydroxylases and related enzymes, Crit. Rev. Biochem. Mol. Biol. 39 (2004) 21–68.
100. O. Sundheim, C.B. Vagbø, M. Bjør ˚ as, M.M.L. Sousa, V. ˚ Talstad, P.A. Aas, F. Drabløs, H.E. Krokan, J.A. Tainer, G.

Slupphaug, Human ABH3 structure and key residues for oxidative demethylation to reverse DNA/RNA damage, EMBO J. 25 (2006) 3389–3397.

101. Y. Kubota, R.A. Nash, A. Klungland, et al., Reconstitution of DNA base excision-repair with purified human proteins: interaction between DNA polymerase beta and the XRCC1 protein, EMBO J. 15 (1996) 6662–6670.
102. Z.O. Jonsson, R. Hindges, U. H ̈ubscher, Regulation of DNA replication and repair proteins through interaction with the front side of proliferating cell nuclear antigen, EMBO J. 17 (1998) 2412–2425.
103. K.Y. Hwang, K. Baek, H.-Y. Kim, Y. Cho, The crystal structure of flap endonuclease-1 from Methanococcus jannaschii, Nat. Struct. Biol. 5 (1998) 707–713.
104. B. Chapados, D. Hosfield, S. Han, J. Qiu, B. Yelent, B. Shen, J. Tainer, Structural basis for FEN-1 substrate specificity and PCNA-mediated activation in DNA replication and repair, Cell 116 (2004) 39–50.
105. S. Sakurai, K. Kitano, H. Yamaguchi, K. Hamada, K. Okada, K. Fukuda, M. Uchida, E. Ohtsuka, H. Morioka, T. Hakoshima, Structural basis for recruitment of human flap endonuclease 1 to PCNA, EMBO J. 24 (2005) 683– 693.
106. T.S.R. Krishna, X.-P. Kong, S. Gary, P.M. Burgers, J. Kuriyan, Crystal structure of the eukaryotic DNA polymerase processivity factor PCNA, Cell 79 (1994) 1233–1243.
107. R.W. Sobol, J.K. Horton, R. Kuhn, et al., Requirement of mammalian DNA polymerase-beta in base-excision repair, Nature 379 (1996) 183–186.
108. F.D. Jay II, J.A. Robert, H. Zuzana, A.F. Rose, H. Zdenek, 2.3 A crystal structure of the catalytic domain of DNA polymerase beta, Cell 76 (1994) 1123–1133.
109. H. Pelletier, M.R. Sawaya, A. Kumar, S.H. Wilson, J. Kraut, Structures of ternary complexes of rat DNA polymerase beta, a DNA template-primer, and ddCTP, Science 264 (1994) 1891–1903.

110. M.R. Sawaya, H. Pelletier, A. Kumar, S.H. Wilson, J. Kraut, Crystal structure of rat DNA polymerase beta: evidence for a common polymerase mechanism, Science 264 (1994) 1930–1935.
111. H. Pelletier, M.R. Sawaya, W. Wolfle, S.H. Wilson, J. Kraut, Crystal structures of human DNA polymerase beta complexed with DNA: implications for catalytic mechanism, processivity, and fidelity, Biochemistry 35 (1996) 12742–12761.
112. M.R. Sawaya, R. Prasad, S.H. Wilson, J. Kraut, H. Pelletier, Crystal structures of human DNA polymerase beta complexed with gapped and nicked DNA: evidence for an induced fit mechanism, Biochemistry 36 (1997) 11205–11215.
113. J.W. Arndt, W. Gong, X. Zhong, A.K. Showalter, J. Liu, C.A. Dunlap, Z. Lin, C. Paxson, M.D. Tsai, M.K. Chan, Insight into the catalytic mechanism of DNA polymerase beta: structures of intermediate complexes, Biochemistry 40 (2001) 5368–5375.
114. J. Krahn, W. Beard, H. Miller, A. Grollman, S. Wilson, Structure of DNA polymerase beta with the mutagenic DNA lesion 8-oxodeoxyguanine reveals structural insights into its coding potential, Structure 11 (2003) 121–127.
115. J.M. Krahn, W.A. Beard, S.H. Wilson, Structural insights into DNA polymerase beta deterrents for misincorporation support an induced-fit mechanism for fidelity, Structure 12 (2004) 1823–1832.
116. S. Johnson, L. Beese, Structures of mismatch replication errors observed in a DNA polymerase, Cell 116 (2004) 803–816.
117. V.K. Batra, W.A. Beard, D.D. Shock, L.C. Pedersen, H.W. Samuel, Nucleotide-induced DNA polymerase active site motions accommodating a mutagenic DNA intermediate, Structure 13 (2005) 1225–1233.
118. Y. Matsumoto, K. Kim, Excision of deoxyribose phosphate residues by DNA polymerase beta during DNA repair, Science 269 (1995) 699–702.

119. R. Prasad, V.K. Batra, X.-P. Yang, J.M. Krahn, L.C. Pedersen, W.A. Beard, S.H. Wilson, Structural insight into the DNA polymerase P deoxyribose phosphate lyase mechanism, DNA repair (Amst) 4 (2005) 1347–1357.
120. A.E. Tomkinson, S. Vijayakumar, J.M. Pascal, T. Ellenberger, DNA ligases: structure, reaction mechanism, and function, Chem. Rev. 106 (2006) 687–699.
121. Z.B. Mackey, C. Niedergang, J.M. Murcia, J. Leppard, K. Au, J. Chen, G. Murcia, A.E. Tomkinson, DNA ligase III is recruited to DNA strand breaks by a zinc finger motif homologous to that of poly(ADP-ribose) polymerase, J. Biol. Chem. 274 (1999) 21679–21687.
122. A.W. Kulczyk, J.C. Yang, D. Neuhaus, Solution structure and DNA binding of the zinc-finger domain from DNA ligase III, J. Mol. Biol. 341 (2004) 723–738.
123. M.S. Satoh, T. Lindahl, Role of poly (ADP-ribose) formation in DNA repair, Nature 356 (1992) 356–358.
124. V.V. Krishnan, K.H. Thornton, M.P. Thelen, M. Cosman, Solution structure and backbone dynamics of the human DNA ligase III BRCT domain, Biochemistry 40 (2001) 13158–13166.
125. A.E. Tomkinson, D.S. Levin, Mammalian DNA ligase, BioEssay 19 (1997) 893–901.
126. W. Ramos, N. Tappe, J. Talamantez, E.C. Friedberg, A.E. Tomkinson, Two distinct DNA ligase activities in mitotic extracts of the yeast Saccharomyces cerevisiae, Nucleic Acids Res. 25 (1997) 1485–1492.
127. X. Zhang, S. Morera, P.A. Bates, P.C. Whitehead, A.I. Coffer, K. Hainbucher, et al., Structure of an XRCC1 BRCT domain: a new protein–protein interaction module, EMBO J. 17 (1998) 6404–6411.
128. J.M. Pascal, P.J. O'Brien, A.E. Tomkinson, T. Ellenberger, Human DNA ligase I completely encircles and partially unwinds nicked DNA, Nature 432 (2004) 473–478.

129. H. Nishida, S. Kiyonari, Y. Ishino, K. Morikawa, The closed structure of an archaeal DNA ligase from Pyrococcus furiosus, J. Mol. Biol. 360 (2006) 956–967.
130. C.D. Mol, A.S. Arvai, R.J. Sanderson, G. Slupphaug, B. Kavli, H.E. Krokan, D.W. Mosbaugh, J.A. Tainer, Crystal structure of human uracil-DNA glycosylase in complex with a protein inhibitor: protein mimicry of DNA, Cell 82 (1995) 701–708.
131. C.D. Putnam, M.J.N. Shroyer, A.J. Lundquist, C.D. Mol, A.S. Arvai, D.W. Mosbaugh, J.A. Tainer, Protein mimicry of DNA from crystal structures of the uracil-DNA glycosylase inhibitor protein and its complex with Escherichia coli uracil-DNA glycosylase, J. Mol. Biol. 287 (1999) 331–346.
132. C.D. Putnam, J.A. Tainer, Protein mimicry of DNA and pathway regulation, DNA Repair (Amst) 4 (2005) 1410–1420.
133. A. Sarker, S. Tsutakawa, S. Kostek, C. Ng, D. Shin, M. Peris, E. Campeau, J. Tainer, E. Nogales, P. Cooper, Recognition of RNA polymerase II and transcription bubbles by XPG, CSB, and TFIIH: insights for transcription-coupled repair and Cockayne Syndrome, Mol. Cell 20 (2005) 187–198.
134. C.D. Putnam, S.B. Clancy, H. Tsuruta, S. Gonzalez, J.G. Wetmur, J.A. Tainer, Structure and mechanism of the RuvB Holliday junction branch migration motor, J. Mol. Biol. 311 (2001) 297–310.
135. K. Hopfner, A. Karcher, D. Shin, L. Craig, L. Arthur, J. Carney, J. Tainer, Structural biology of Rad50 ATPase: ATP-driven conformational control in DNA double-strand break repair and the ABC-ATPase superfamily, Cell 101 (2000) 789–800.
136. R.S. Williams, J.A. Tainer, A nanomachine for making ends meet: MRN is a flexing scaffold for the repair of DNA double-strand breaks, Mol. Cell 19 (2005) 724–726.

137. D.S. Shin, L. Pellegrini, D.S. Daniels, B. Yelent, et al., Full-length archaeal Rad51 structure and mutants: mechanisms for RAD51 assembly and control by BRCA2, EMBO J. 22 (2003) 4566–4576.
138. M. Otterlei, B. Kavli, R. Standal, C. Skjelbred, S. Bharati, H.E. Krokan, Repair of chromosomal abasic sites in vivo involves at least three different repair pathways, EMBO J. 19 (2000) 5542–5551.
139. E.C. Chao, S.M. Lipkin, Molecular models for the tissue specificity of DNA mismatch repair-deficient carcinogenesis, Nucleic Acids Res. 34 (2006) 840–852.
140. J. de Boer, J.H. Hoeijmakers, Nucleotide excision repair and human syndromes, Carcinogenesis 21 (2000) 453–460.
141. J.J. Perry, L. Fan, J.A. Tainer, Developing master keys to brain pathology, cancer and aging from the structural biology of proteins controlling reactive oxygen species and DNA repair, Neuroscience (2006) Epub ahead of print.
142. Y. Doi, A. Katafuchi, Y. Fujiwara, K. Hitomi, J.A. Tainer, H. Ide, S. Iwai, Synthesis and characterization of oligonucleotides containing 2 -fluorinated thymidine glycol as inhibitors of the endonuclease III reaction, Nucleic Acids Res. 34 (2006) 1540–1551.
143. J.M. Pascal, O.V. Tsodikov, G.L. Hura, W. Song, E.A. Cotner, S. Classen, A.E. Tomkinson, J.A. Tainer, T. Ellenberger, A flexible interface between DNA ligase and PCNA supports conformational switching and efficient ligation of DNA, Mol. Cell 24 (2006) 279–291.
144. S. Vijayakumar, B.R. Chapados, K.H. Schmidt, R.D. Kolodner, J.A. Tainer, A.E. Tomkinson, The C-terminal domain of yeast PCNA is required for physical and functional interactions with Cdc9 DNA ligase, Nucleic Acids Res. (2007) in press.

145. L. Fan, S. Kim, C.L. Farr, K.T. Schaefer, K.M. Randolph, J.A. Tainer, L.S. Kaguni, A novel processive mechanism for DNA synthesis revealed by structure, modeling and mutagenesis of the accessory subunit of human mitochondrial DNA polymerase, J. Mol. Biol. 358 (2006) 1229–1243.
146. I. Ivanov, B.R. Chapados, J.A. McCammon, J.A. Tainer, Proliferating cell nuclear antigen loaded onto double-stranded DNA: dynamics, minor groove interactions and functional implications, Nucleic Acids Res. 34 (2006) 6023–6033.

Chapter 6

An Approach to Performance Assessment and Fault Diagnosis for Rotating Machinery Equipment

Xiaochuang Tao[1], Chen Lu[1,2], Chuan Lu[1,2], and Zili Wang[1,2]

[1]School of Reliability and Systems Engineering, Beihang University, Beijing 100191, People's Republic of China

[2]Science & Technology Laboratory on Reliability & Environmental Engineering, Beijing 100191, People's Republic of China

ABSTRACT

Predict and prevent maintenance is routinely carried out. However, how to address the problem of performance assessment maximizing the use of available monitoring data, and how to build a framework that integrates performance assessment, fault detection, and diagnosis are still a significant challenge. For this purpose, this article introduces an approach to performance assessment and fault diagnosis for rotating machinery, including wavelet packet decomposition for extracting energy feature samples from vibration signals acquired during normal and faulty conditions; clustering analysis for demonstrating the separability of the samples; and Fisher discriminant analysis for providing an optimal lower-dimensional representation, in terms of maximizing the separability among different populations, by projecting the samples into a new space. In the new low-dimensional space, the Mahalanobis distance (MD) between the new measurement data and normal population can be calculated for performance assessment. Moreover, this model for performance assessment only requires data to be available in normal conditions and any one of all possible fault conditions, without the necessity for the full life cycle of condition monitoring data. In addition, if monitoring data under different fault conditions are available, the fault mode can be identified accurately by comparing the MDs between the new measurement data and each fault population. Finally, the proposed method was verified to be successful on performance assessment and fault diagnosis via a hydraulic pump test and a ball bearing test.

INTRODUCTION

Currently, driven by the demand to reduce maintenance costs, shorten repair time, and maintain high availability of equipment, maintenance strategies have progressed from breakdown maintenance (fail and fix) to preventive maintenance, then to condition-based maintenance (CBM), and lately toward a prospect of intelligent predictive maintenance (predict and prevent), [1-3]. While the reactively breakdown maintenance and blindly preventive maintenance do sometimes reduce equipment failures, they are more labor intensive, do not eliminate catastrophic failures and cause unnecessary maintenance. This is where CBM steps in. It was reported that 99% of mechanical failures especially rotating machinery are preceded by noticeable indicators [4]. That is to say, with the exception of abrupt, catastrophic failures, most faults of rotating machinery equipment have progression processes to failure. We can view the deterioration as a two-stage process: the first stage as normal operation, and the second stage as a potential failure [5,6]. Of interest here is when the second stage starts and how it develops. CBM attempts to monitor machinery health based on condition measurements that do not interrupt normal machine operation. Rotating machinery is one of the most common classes of machines. Over the past few years, technologies in condition monitoring, fault diagnostics, and prognosis for rotating machinery, which are important aspects in a CBM program, have been receiving more attention. Because fault diagnosis problems can be considered as classification problems [7], Fisher discriminant analysis (FDA), which is studied in detail in the pattern classification literature, has been applied to conduct fault diagnosis. However, at present, its application has mainly been concentrated to industrial process (especially chemical processes), but rarely to rotating machinery equipment [8-11]. Moreover, although performance assessment, fault detection, diagnosis, and prognosis have received increased attention with significant progress [12], currently, few methods can realize those purposes alone. In addition, the incomplete data (monitoring data under normal or fault conditions) have rarely been considered as input data to performance assessment and this has yet to be fully utilized. To fill the gaps, this article uses the combination of FDA and Mahalanobis distance (MD) applied to rotating machinery fault diagnosis, which is further extended to performance assessment and fault detection. Thus, a framework integrating performance assessment, fault detection, and diagnosis is built, which not only solves the problem of when the second stage of potential failure starts and how it develops, but because of its data-driven property also shows appropriate potential and provides a functional interface to performance trend prognosis.

Successful clustering can demonstrate the separability of samples and provide preconditions for FDA. To avoid the influence of the dispersibility of sample data acquired during normal and various fault conditions, the analysis of the separability of sample data is indispensable. Cluster analysis can classify samples into corresponding groups based on the measured parameters, and hierarchical cluster analysis (HCA) is the most commonly used clustering tool [13,14]. FDA provides an optimal lower dimensional representation, in terms of maximizing the separability among different populations representative of different operational states, by projecting normal and fault populations, and separating them to the limit in the reconstructed space [15-17]. In the reconstructed low-dimensional space, the MD between the new measurement data and the normal population, constructed using normal data, can be calculated for performance assessment. The MD can also be transformed into a normalized confidence value (CV), according to the presupposed threshold. If an abnormal state is detected by performance assessment, the MDs between the new measurement data and the normal and different fault populations are calculated, to identify which population the new data belong to, and thus, the fault mode can be recognized.

The proposed method for performance assessment only needs monitoring data under normal conditions and any one of all possible fault conditions. As for fault diagnosis, if the monitoring data under different fault conditions are available, accurate diagnosis results can be achieved by this model. Moreover, in this model, the algorithm is simple and intuitive and offers good interpretation for the results. The proposed method was also verified to be effective and pragmatic for performance assessment and fault diagnosis via a hydraulic pump test and a ball bearing test.

METHODOLOGY FOR PERFORMANCE ASSESSMENT AND FAULT DIAGNOSIS

Wavelet Packet Decomposition-based Feature Extraction

In practice, the characteristic frequencies of rotating machinery equipment or components are usually distributed in both high- and low-frequency bands. In view of this fact, wavelet packet analysis (WPA) is proposed to construct a more sophisticated method of orthogonal decomposition based on multi-resolution analysis, which can divide the full frequency band into multi-levels, so that each band contains information that is more specific [18,19]. Therefore, wavelet packet decomposition is suitable for extracting both low- and high-frequency features. Statistically analyzing all bands of a signal decomposed by wavelet packet, an energy index of each frequency band can be extracted.

The determination of the wavelet packet decomposition scale is an issue that cannot be ignored. If the decomposition scale is too little, the fault features cannot effectively be extracted, whereas too many scales will increase the dimension of the feature vector, and consequently, the calculating rate can be affected [20,21]. Therefore, in the hydraulic pump and ball bearing performance assessments, according to the frequency spectrum analysis of the vibration signal explained in Section 3, eight frequency band energy indexes E_3j can be calculated by three-layer decomposition.

$$E_{3j} = \int |S_{3j}(t)|^2 dt = \sum_{k=1}^{n} |x_{jk}|^2 \quad (1)$$

where x_{jk} ($j = 0, 1, \ldots, 7; k = 1, 2, \ldots, n$) denotes the amplitude of discrete points in the reconstructed signal S_{3j}.

When a hydraulic pump or bearing progresses into a state of degradation, the energy of each frequency band will have a great impact, and therefore, the energy can be normalized into a feature vector T.

$$T=[E_{30}/E, E_{31}/E, E_{32}/E, E_{33}/E, E_{34}/E, E_{35}/E, E_{36}/E, E_{37}/E] \tag{2}$$

$$E = \left(\sum_{j=0}^{7} |E_{3j}|^2 \right)^{1/2} \tag{3}$$

Separability Analysis of Sample Dataseparability Analysis of Sample Data

In order to avoid the influence of the dispersibility of sample data acquired during normal and various fault conditions, analysis of the separability of the sample data is indispensable.

Clustering is an unsupervised pattern classification method, in which the goal is to determine a finite set of categories to describe a dataset according to similarities among its objects [22,23]. As one of the most popular clustering methods, HCA consists of mathematically treating each sample as a point in multidimensional space described by the chosen variables; it builds a nested partition set called a cluster hierarchy. When a given sample is taken as a point in the space defined by the variables, the distance between this point and all the other points can be calculated, thereby establishing a matrix that describes the proximity between all the samples studied. Based on this matrix of proximity between the samples, one can construct a similarity diagram called a dendrogram. There are many ways of mathematically grouping these points in multidimensional space in order to form hierarchical clusters [24-26].

FDA

As an optimal linear dimensionality reduction technique, in terms of maximizing the separation between different populations, FDA has been studied in detail in the pattern classification literature[27-29].

For either performance assessment or fault diagnosis, data collected from the unit during normal and various fault states are categorized into different populations, where each population contains data representing a particular state.

Definition (MD)

Given the covariance matrix of a *p*-variate population G as Σ ($\Sigma > 0$), and x, y are two samples taken from G. Defining

$$d^2(x, y) = (x - y)' \Sigma^{-1} (x - y) \tag{4}$$

Then $d(x,y)$ is called the MD between x and y. Defining

$$d^2(x, G) = (x - \mu)' \Sigma^{-1} (x - \mu) \tag{5}$$

where μ is the mean vector of G, then $d(x,G)$ is called the MD between x and the population G.

Lemma

Given A is a *p*-order symmetric matrix, and $B > 0$ is a *p*-order positive definite matrix, the eigenvalues of $B^{-1}A$ are $\lambda_1 \geq \lambda_2 \geq \cdots \geq \lambda_p$ and the corresponding standard eigenvectors are $a_1, a_2, \ldots, a_p$ (standardized into $a_i' B a_i = 1$). Then, the optimization problem is described as below.

$$\begin{cases} x'Ax \to \max \\ x'Bx = 1 \end{cases} \tag{6}$$

when $x = a_1$, the maximum value can be achieved as λ_1.

$$(P_k) \quad \begin{cases} x'Ax \to \max \\ x'Bx = 1, x'Ba_i = 0, \\ i = 1, 2, \ldots, k-1 \end{cases} \tag{7}$$

when $x = a_k$, the maximum value can be achieved as λ_k.

For a more detailed discussion of the extremal problem of the quadratic form, refer to [30].

The mathematical derivation process of FDA follows.

Define $G_1 \sim (\mu_1, \Sigma_1), G_2 \sim (\mu_2, \Sigma_2), \ldots, G_k \sim (\mu_k, \Sigma_k)$ as k populations, where μ_i and Σ_i are, respectively, the mean vector and covariance matrix of G_i. $x \in R^p$ is a sample to be determined. Through a linear combination of variable indexes in each population, corresponding one-dimensional sample $y = a'x$ can be achieved, which may come from any one of those populations $G_1^* \sim (a'\mu_1, a'\Sigma_1 a)$, $G_2^* \sim (a'\mu_2, a'\Sigma_2 a)$, ..., $Gk^* \sim (a'\mu_k, a'\Sigma_k a)$. Then, defining B_0 and E_0 as below

$$B_0 = \sum_{i=1}^{k} n_i(a'\mu_i - a'\bar{\mu})^2$$

$$= a'\left[\sum_{i=1}^{k} n_i(\mu_i - \bar{\mu})(\mu_i - \bar{\mu})'\right]a = a'Ba \tag{8}$$

$$E_0 = \sum_{i=1}^{k} a' \sum_i a = a'(\Sigma_1 + \Sigma_2 + \ldots + \Sigma_k)a = a'Ea \tag{9}$$

where is the number of samples for Gi, n is the total number of samples. Then B is the between-class-scatter matrix, and E is the within-class-scatter matrix.

Thinking along the lines of variance analysis, in order to better separate each population, the choice of a should make B_0 expand as far as possible, while E_0 narrows as far as possible.

Thus, the first FDA vector a can be determined as

$$\max_a \left(\frac{B_0}{E_0} = \frac{a'Ba}{a'Ea}\right) \tag{10}$$

This is equivalent to the following optimization problem.

$$(P) \begin{cases} a'Ba \to \max \\ a'Ea = 1 \end{cases} \tag{11}$$

According to the lemma given in Section 2.3.2, when a is the standard eigenvector a_1 (standardized into $a_i'Ba_i = 1$) corresponding to the maximum eigenvalue λ_1 of $E^{-1}B$, formula (10) achieves the maximum λ_1. Thus, the first canonical variable is $y_1 = a_1'x$.

The second FDA vector is computed to maximize the scatter between classes, while minimizing the scatter within classes, among all axes perpendicular to the first FDA vector a_1. According to the lemma given in Section 2.3.2, the second canonical variable is $y_2 = a_2'x$ and so on for the remaining FDA vectors and canonical variables. Usually, given the first m eigenvalues as $\lambda_1 \geq \lambda_2 \geq \cdots \geq \lambda_m$, the corresponding standard eigenvectors as $a_1, a_2, \ldots, am$, when the accumulation contributionreaches a threshold (such as 85%), we can get m unrelated canonical variables $y_1, y_2, \ldots, y_m, y_m = a_m'x$ for discriminant analysis. This is equivalent to mapping a variable from p-dimensional space to m-dimensional space for analysis, where $m < p$.

After dimensionality reduction, according to the definition given in Section 2.3.1, the MD $d(x, G_j)$ between x and G_j can be achieved by calculating the distance between $y = (y_1, y_2, \ldots, y_m)'$ and $G_j^*(j = 1, 2, \ldots, k)$. The logic process of FDA is shown as Figure 1.

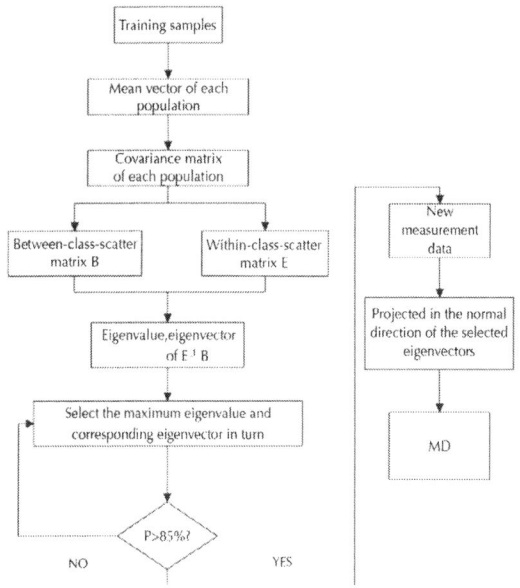

Figure 1: Logic diagram of FDA.

MD for Performance Assessment

In practice, the whole life cycle of condition-monitoring data acquired from a machine is already scarce due to irregular measurement recording, and/or the huge amount of time they take to accumulate. For example, a bearing may last several years even under harsh operating conditions. Therefore, regardless of whether it is an experimental or practical application, it is hard to acquire condition-monitoring data that are representative of the whole life cycle. More commonly, what we can get are incomplete data, such as normal data and fault data [3,31]. Usually, the incomplete data are rarely to be considered as input data of performance assessment, and this has yet to be fully utilized. Consequently, maximizing the use of available data to address the problem of performance assessment is a significant challenge.

As previously mentioned, even if only datasets under normal conditions and any one of all possible fault conditions are available, the normal population can still accurately be clustered and characterized. As shown in Figure 2, using FDA, a space conversion method can be achieved in which, the normal population and new measurement data can be projected from the original high-dimensional space into a new low-dimensional space. Thus, the issue on performance assessment can be addressed by the MD away from the normal population [32]. The MD is calculated between the feature vector extracted from online monitoring data and the normal population and thus, the MD, which indicates how far the input data deviate from the region of normal conditions, can reveal the current performance state. If it exceeds the predetermined threshold (the threshold value can be determined based on engineering experience), the process is probably in an abnormal state. The MD can be defined as

$$MD = d(x, G_{normal}) \qquad (12)$$

where x is an input feature vector, and G_{normal} is the normal population.

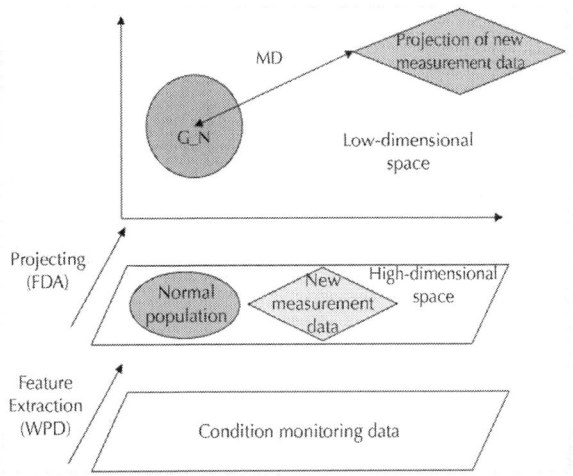

Figure 2: FDA map for performance assessment.

The performance can be quantized and visualized by following the novel pattern of MD. However, the MD is only an absolute index, without a reference index, it would be difficult to determine whether the current performance condition was good or bad. To succinctly describe the current performance state, the MD coupling with a benchmark can be transformed into a normalized CV, ranging from 0 to 1. A higher CV closer to 1 indicates a performance state closer to normal, while a lower CV closer to 0 is closer to a condition of failure.

$$CV = e^{-\frac{\sqrt{MD}}{c}} \qquad (13)$$

where c is a scale parameter, which is determined by the averaged MDs under normal state and a predetermined CV benchmark.

In summary, the process of performance assessment is shown in Figure 3.

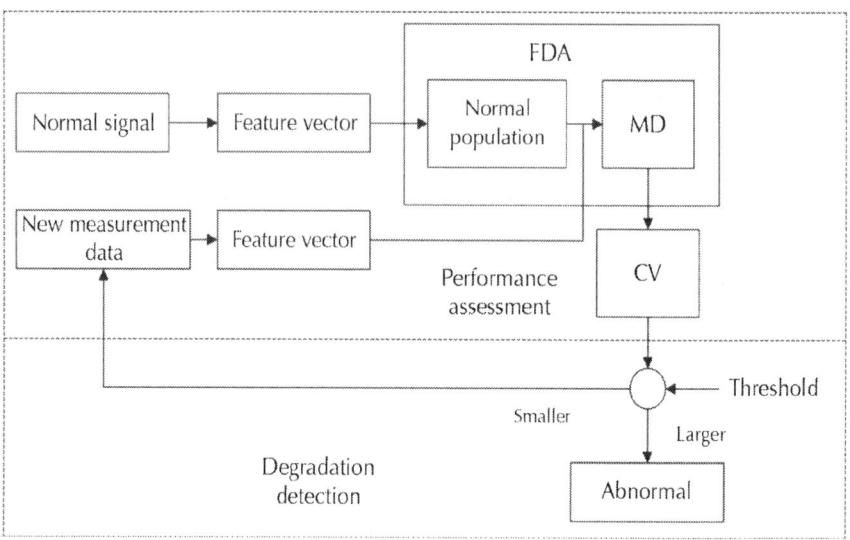

Figure 3: Performance assessment process based on FDA & MD.

MD for Fault Diagnosis

If monitoring data under different fault conditions are available, several groups of feature vector samples can be extracted as a learning set. As shown in Figures 4 and 5, with HCA, the learning set will be classified into different populations corresponding to the original operational states. On the basis of demonstration of the separability of feature vector samples, those clustered populations representative of different states, including normal and faults, are analyzed by FDA. FDA provides an optimal lower-dimensional representation in terms of maximizing the separability among different populations. In the new low-dimensional space, the MDs between the real measurement data and those different populations can be calculated, according to the discriminant rule as follows

An Approach to Performance Assessment and Fault Diagnosis...

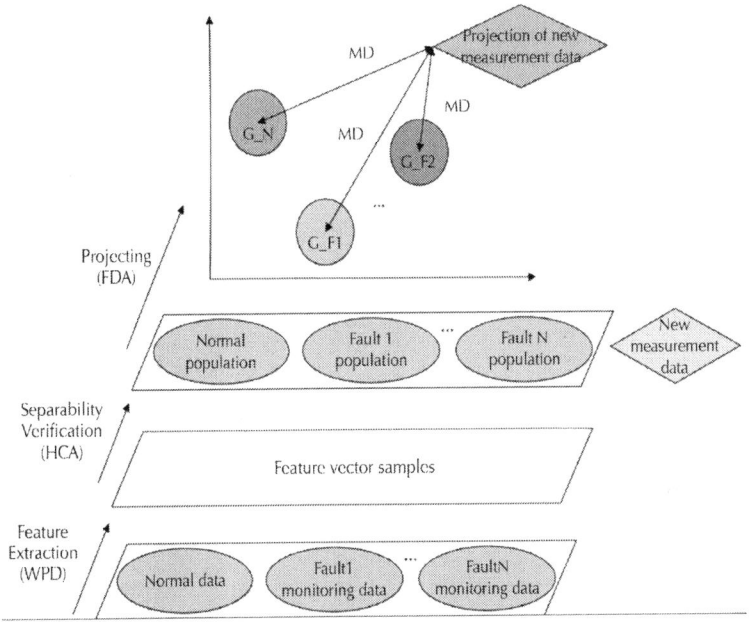

Figure 4: FDA map for fault diagnosis.

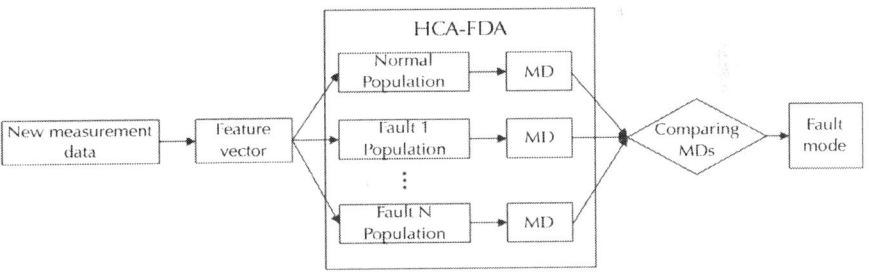

Figure 5: Fault diagnosis process based on FDA & MD.

If $d^2(x,G_l) = \min_{1 \leq j \leq k} d2(x,G_j)$, then $x \in G_l$.

The state represented by the population that has the minimum MD with the real measurement data can be identified as the current operating state and thus fault diagnosis is completed.

EXPERIMENTAL VERIFICATION

Two experimental cases (i) hydraulic pump performance assessment and fault diagnosisand (ii) ball bearing performance assessment and fault diagnosis are presented to validate the effectiveness and practicality of applying the proposed method to performance assessment and fault diagnosis of rotating machinery equipment. The description of these case studies will follow the sequence: (1) experimental setup and data acquisition; (2) signal analysis and feature extraction; (3) analysis of performance assessment and fault diagnosis results.

Performance Assessment and Fault Diagnosis for Hydraulic Pump

The hydraulic pump is the heart of a hydraulic system, which determines whether the whole system can run normally or not. Therefore, performance assessment and fault diagnosis of the hydraulic pump is of great importance. Usually, when a hydraulic pump is under an abnormal state, it will be revealed by changes of vibration. Because most mechanical faults are reflected by vibration [33], the vibration signals of the hydraulic pump are collected and analyzed in this experiment for performance assessment and fault diagnosis.

Experimental Setup and Data Acquisition

In this experiment, a hydraulic pump was tested and analyzed, as shown in Figure 6. Two commonly occurring faults in hydraulic pumps were set: slipper loose and valve plate wear. Under different states (normal, faults), monitoring data (vibration signal) were, respectively, acquired from the hydraulic pump end using an acceleration transducer, with motor speed stabilized at 528 rpm, and a sampling frequency of 1000 Hz.

Figure 6: Test-rig of hydraulic pump.

Feature Extraction by WPA

To determine the wavelet packet decomposition scale, FFT was implemented for the vibration signals acquired under normal, slipper loose, and valve plate wear states. Through the analysis of the frequency spectrum, as shown in Figures 7 and 8, it can be seen that there are clear peaks appearing evenly at six characteristic frequencies. Furthermore, the normal signal and fault (slipper loose) signal have distinctively different peaks at the six characteristic frequencies. Therefore, each set of acquired vibration signals is suitable to be decomposed into eight frequency bands by three-layer wavelet packet decomposition. In that way, all the peaks can be contained in different frequency bands. Then, an eight-dimensional feature vector can be constructed by calculating and normalizing the energy of each band. The frequency range corresponding to each frequency band is $(n\omega_{max}/2^N, (n + 1)\omega_{max}/2^N)$, where $N = 3$, $n = 0, 1,...,7$, and ω_{max} is the maximum frequency; here $\omega_{max} = 498$.

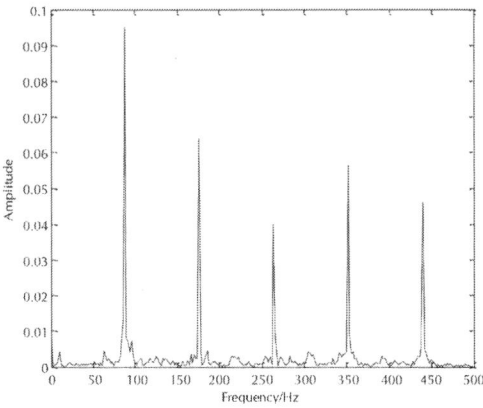

Figure 7: FFT spectrum of acquired normal signal.

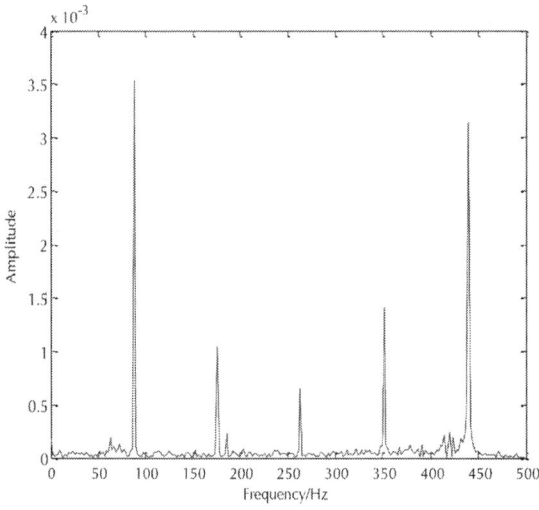

Figure 8: FFT spectrum of acquired fault (slipper loose) signal.

For those three states, eight feature vector samples were acquired, respectively. The first four samples of each state were used as the FDA learning set (for clearer identification they were numbered as Normal_1 to Normal_4, Fault1_1 to Fault1_4, and Fault2_1 to Fault2_4, corresponding to the normal state, slipper loose state, and valve plate wear conditions, respectively), while the others were used as the testing set (they were also numbered in the same manner as: N_1 to N_4, F1_1 to F1_4, and F2_1 to F2_4, respectively).

The feature vector samples for learning and testing are shown in Tables 1 and 2.

Table 1: Feature vector samples for learning (hydraulic pump)

Number	1	2	3	4	5	6	7	8
Normal_1	0.0304	0.8484	0.1399	0.3570	0.1182	0.1565	0.1143	0.2843
Normal_2	0.0343	0.8433	0.1396	0.3557	0.1061	0.1819	0.1085	0.2926
Normal_3	0.0305	0.8307	0.1360	0.3732	0.1231	0.1659	0.1228	0.3057
Normal_4	0.0280	0.8163	0.1654	0.3778	0.1183	0.1491	0.1432	0.3256
Fault1_1	0.0256	0.8182	0.0260	0.1312	0.3582	0.4088	0.0290	0.1255
Fault1_2	0.0329	0.8141	0.0310	0.1312	0.3450	0.4265	0.0355	0.1255
Fault1_3	0.0355	0.8256	0.0378	0.1357	0.3483	0.3972	0.0323	0.1309
Fault1_4	0.0329	0.8266	0.0358	0.1284	0.3950	0.3517	0.0341	0.1304
Fault2_1	0.0261	0.2428	0.5679	0.0871	0.0190	0.0361	0.7785	0.0507
Fault2_2	0.0156	0.2385	0.5717	0.0861	0.0229	0.0366	0.7775	0.0484
Fault2_3	0.0193	0.2405	0.5628	0.0862	0.0278	0.0283	0.7834	0.0488
Fault2_4	0.0116	0.2479	0.5393	0.0816	0.0275	0.0319	0.7977	0.0521

Tao et al.
Tao et al. EURASIP Journal on Advances in Signal Processing 2013 2013:5, doi: 10.1186/1687-6180-2013-5

Table 2: Feature vector samples for testing (hydraulic pump)

Number	1	2	3	4	5	6	7	8
N_1	0.0428	0.8110	0.1647	0.4138	0.1297	0.1359	0.1495	0.2908
N_2	0.0300	0.8195	0.1626	0.4070	0.1180	0.1437	0.1434	0.2833
N_3	0.0353	0.8078	0.1836	0.4182	0.1089	0.1534	0.1464	0.2841
N_4	0.0326	0.8142	0.1716	0.4177	0.1107	0.1467	0.1538	0.2732
F1_1	0.0293	0.8259	0.0367	0.1255	0.4254	0.3138	0.0316	0.1397
F1_2	0.0243	0.8344	0.0266	0.1398	0.3832	0.3446	0.0229	0.1295
F1_3	0.0318	0.8399	0.0281	0.1274	0.3242	0.3902	0.0266	0.1356
F1_4	0.0360	0.8293	0.0338	0.1308	0.4289	0.2984	0.0237	0.1383
F2_1	0.0088	0.2475	0.5249	0.0839	0.0189	0.0374	0.8075	0.0484
F2_2	0.0185	0.2501	0.5710	0.0907	0.0299	0.0232	0.7739	0.0511
F2_3	0.0095	0.2463	0.5446	0.0841	0.0256	0.0263	0.7945	0.0541
F2_4	0.0102	0.2476	0.5235	0.0883	0.0236	0.0308	0.8079	0.0513

Tao et al.

Tao et al. EURASIP Journal on Advances in Signal Processing 2013 2013:5, doi: 10.1186/1687-6180-2013-5

Analysis of States Separability

In order to demonstrate the separability of feature vector samples acquired during normal, slipper loose, and valve plate wear states, HCA was selected and carried out on the learning set. As shown in Figure 9, the tree dendrogram clearly shows the whole process of clustering. First, each sample was taken as a class. After the first clustering, according to the distance between classes, samples Fault2_1 to Fault2_4 were merged into a cluster, samples Normal_1 to Normal_4 were merged into another cluster, and samples Fault1_1 to Fault1_4 were merged into the third cluster. Moreover, from the cluster membership, as shown in Table 3, it was found that all the normal samples gathered in cluster 1, all the slipper loose samples gathered in cluster 2, and all the valve plate wear samples gathered in cluster 3. The results are consistent perfectly with the practical situation, and can be seen as strong evidence for the separability of the different states.

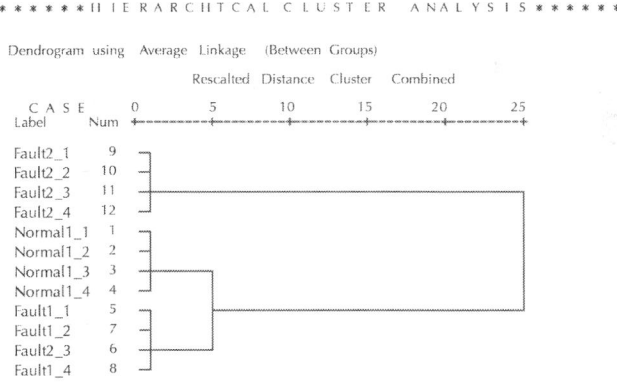

Figure 9: A tree dendrogram of HCA (hydraulic pump).

Table 3: Clustering membership of learning set (hydraulic pump)

Case	Cluster index
1:Normal_1	1

2:Normal_2	1
3:Normal_3	1
4:Normal_4	1
5:Fault1_1	2
6:Fault1_2	2
7:Fault1_3	2
8:Fault1_4	2
9:Fault2_1	3
10:Fault2_2	3
11:Fault2_3	3
12:Fault2_4	3

Tao et al.

Tao et al. EURASIP Journal on Advances in Signal Processing 2013 2013:5, doi: 10.1186/1687-6180-2013-5

Analysis of Performance Assessment Results

In this study, for the purpose of performance assessment, the normal samples numbered as Normal_1 to Normal_4 were used to construct and characterize the normal population noted as G_N. Through FDA, a space conversion method was achieved, by which, the different populations and the new measurement data can be projected from the original eight-dimensional space into a new two-dimensional space.

Then, the MDs between the normal population G_N and those samples included in the testing set, as shown in Table 2, were calculated. They were transformed into normalized CVs according to formula (13). The MD and CV curves are shown in Figures 10 and 11. Obviously, contrasting with the MDs of the normal testing samples, the MDs of the slipper loose and valve plate wear testing samples were quite large, because the normal testing samples were located nearby the normal population G_N, while the fault testing samples were located far away, i.e., the slipper loose and valve plate wear samples were in an abnormal condition. Conversely, the CVs of the normal testing samples were relatively high; close to 0.9. Whereas the CVs of the fault testing samples were all quite low; falling below the presupposed threshold of 0.6. Therefore, the CV index indicated that the normal testing samples were in a normal condition; while the slipper loose and valve plate wear testing samples were in a faulty condition. The analysis demonstrated that the performance assessment could be quantized and visualized by MD and CV, and coupled with a presupposed threshold where abnormal states can be detected. Therefore, this is a successful trial of the performance assessment and fault detection.

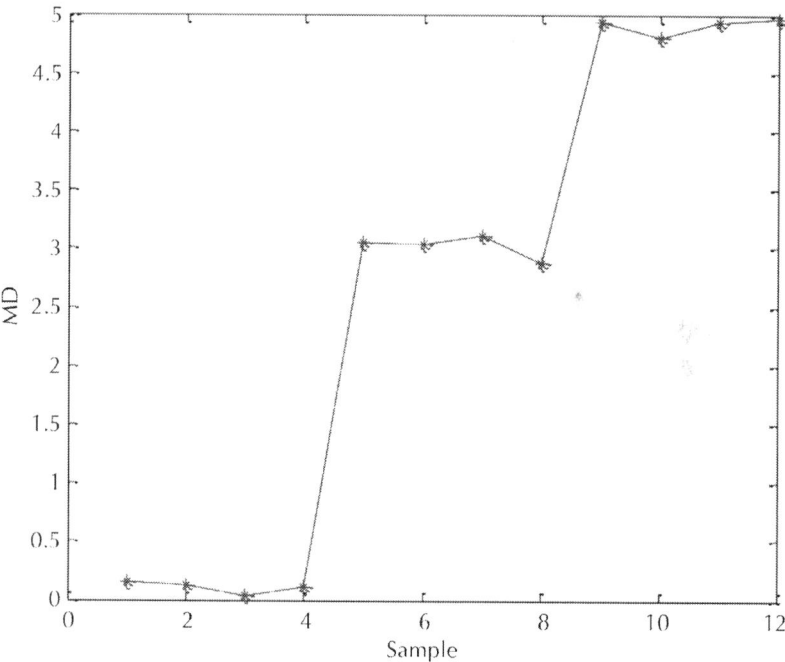

Figure 10: MD result of testing set (hydraulic pump).

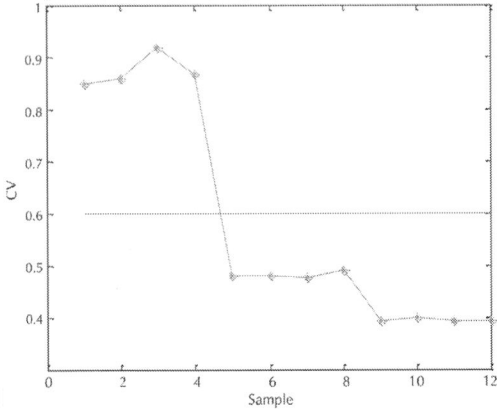

Figure 11: CV result of testing set (hydraulic pump).

Analysis of Fault Diagnosis Results

As previously mentioned in the performance assessment analysis, the samples in the testing set (numbered as F1_1 to F1_4 and F2_1 to F2_4) were detected as fault states. To identify which type of fault they belonged to, as a reference, the normal samples in the testing set (numbered as N_1 to N_4) were also considered. The MDs between those samples in the testing set and these three populations were calculated, as shown in Table 4. In order to facilitate analysis, the normal population was noted as G_N, the slipper loose and valve plate wear state populations were noted as G_F1 and G_F2, respectively. Through comparative analysis, the samples under normal conditions had the smallest MDs with the population G_N, while the samples under slipper loose condition and valve plate wear condition had the smallest MDs with the population G_F1 and G_F2, respectively, as shown in the 'Min' row of Table 4. According to the discriminant rule in Section 2.5, it can be determined that the samples N_1 to N_4 belonged to normal conditions, the samples F1_1 to F1_4 and F2_1 to F2_4 were under slipper loose state and valve plate wear state, respectively, as shown in the 'Mode' row. This accurate diagnosis result is further proof of the effectiveness of the proposed method in fault diagnosis.

Table 4: MDs and diagnosis results of samples in testing set (hydraulic pump)

	N_1	N_2	N_3	N_4	F1_1	F1_2	F1_3	F1_4	F2_1	F2_2	F2_3	F2_4
G_N	0.1490	0.1288	0.0413	0.1135	3.0432	3.0394	3.1073	2.8714	4.9359	4.8058	4.9323	4.9614
G_F1	3.5555	3.5412	3.4477	3.5199	0.3838	0.3794	0.3107	0.5563	7.8717	7.7396	7.8727	7.9023
G_F2	4.7123	4.7364	4.8134	4.7444	7.4089	7.4290	7.5040	7.2451	0.0885	0.0495	0.0928	0.1209
Min	0.1490	0.1288	0.0413	0.1135	0.3838	0.3794	0.3107	0.5563	0.0885	0.0495	0.0928	0.1209
Mode	G_N	G_N	G_N	G_N	G_F1	G_F1	G_F1	G_F1	G_F2	G_F2	G_F2	G_F2

Tao et al.

Tao et al. EURASIP Journal on Advances in Signal Processing 2013 **2013**:5, doi: 10.1186/1687-6180-2013-5

Performance Assessment and Fault Diagnosis for Ball Bearing

Bearings are critical components in rotating machines because their failure could lead to serious damage in machines. In recent years, bearing fault diagnosis has received increasing attention [34-36]. In this case, vibration signals are acquired from the ball bearing housing for performance assessment and fault diagnosis.

Experimental Setup and Data Acquisition

In this case study, the test data were acquired from the Case Western Reserve University Bearing Data Center. As shown in Figure 12, the test-rig consists of a 2-horsepower motor (left), a torque transducer/encoder (center), a dynamometer (right), and control electronics (data not shown). The test bearings support the motor shaft. Single point faults were introduced separately at the inner-race, outer-race, and rolling element (i.e., ball) of the test bearing using electro-discharge machining with fault diameters of 7 mm. Faulted bearings were reinstalled into the test motor, and vibration data were recorded under different four states (normal, faults) using accelerometers, with motor loads of 2 horsepower, motor speed of 1750 rpm, and a sampling frequency of 12,000 Hz.

Figure 12: Test-rig of ball bearing.

Feature Extraction by WPA

Through implementation of FFT on the acquired vibration data, it was found that the characteristic frequencies under normal, inner-race fault, outer-race fault, and ball fault states are distributed separately in different frequency bands, and that there is little overlap between any two adjacent frequencies in the spectrum. Therefore, three-layer wavelet packet decomposition can also be applied to the vibration data, and thus, eight-dimensional feature vectors can be acquired by calculating and normalizing the energy of each band.

For those four states, eight feature vector samples were acquired. The first four samples of each state were used as the FDA learning set (for clearer identification they were numbered as N_1 to N_4, I_1 to I_4, O_1 to O_4, and B_1 to B_4, corresponding to the normal state, inner-race fault state, outer-race fault state, and ball fault state, respectively), while the others were used as the testing set (they were also numbered in the same manner as N_T1 to N_T4, I_T1 to I_T4, O_T1 to O_T4, and B_T1 to B_T4, respectively).

The feature vector samples for learning and testing are shown in Tables 5 and 6.

Table 5: Feature vector samples for learning (rolling bearing)

Number	1	2	3	4	5	6	7	8
N_1	0.8111	0.5012	0.0524	0.2964	0.0002	0.0018	0.0053	0.0187
N_2	0.8259	0.4665	0.0544	0.3112	0.0002	0.0019	0.0054	0.02
N_3	0.7891	0.5274	0.0524	0.3097	0.0002	0.0018	0.0053	0.0204
N_4	0.8137	0.493	0.0553	0.3023	0.0002	0.0018	0.0056	0.0194
I_1	0.0795	0.2297	0.5852	0.1497	0.0014	0.0081	0.7446	0.1471
I_2	0.0823	0.2226	0.5831	0.1464	0.0015	0.0082	0.7467	0.157
I_3	0.081	0.2341	0.6031	0.1474	0.0014	0.0084	0.7306	0.1388
I_4	0.0723	0.2279	0.5922	0.1505	0.0014	0.0092	0.7412	0.1417
O_1	0.0069	0.0096	0.4907	0.0156	0.0082	0.0137	0.8681	0.0707
O_2	0.0065	0.0096	0.4945	0.0161	0.0081	0.0137	0.8655	0.0759
O_3	0.0069	0.0098	0.4831	0.0145	0.0076	0.0121	0.8725	0.0696
O_4	0.0065	0.01	0.5217	0.016	0.007	0.0124	0.8492	0.0777
B_1	0.045	0.0449	0.4876	0.0184	0.0005	0.0027	0.8703	0.0181
B_2	0.0443	0.0424	0.457	0.0189	0.0005	0.0025	0.887	0.0179
B_3	0.0426	0.0436	0.464	0.0187	0.0005	0.0025	0.8833	0.0186
B_4	0.0412	0.042	0.4584	0.0179	0.0004	0.0022	0.8865	0.0175

Tao et al.
Tao et al. EURASIP Journal on Advances in Signal Processing 2013 2013:5, doi: 10.1186/1687-6180-2013-5

Table 6: Feature vector samples for testing (rolling bearing)

Number	1	2	3	4	5	6	7	8
N_T1	0.804	0.5112	0.0511	0.2987	0.0002	0.0016	0.0051	0.0192
N_T2	0.8119	0.5007	0.0528	0.2948	0.0002	0.0018	0.0053	0.019
N_T3	0.8208	0.4824	0.0538	0.3006	0.0002	0.0017	0.0055	0.0194
N_T4	0.7854	0.5344	0.0532	0.307	0.0002	0.0018	0.0054	0.0198
I_T1	0.0734	0.2138	0.58	0.1423	0.0015	0.008	0.7556	0.1461
I_T2	0.087	0.2276	0.5929	0.15	0.0014	0.0089	0.7394	0.1403
I_T3	0.0769	0.2242	0.6017	0.1463	0.0016	0.0084	0.7343	0.1454
I_T4	0.071	0.2271	0.5888	0.1535	0.0014	0.0082	0.7421	0.1498
O_T1	0.0066	0.0091	0.4939	0.0159	0.0073	0.0132	0.8661	0.0736
O_T2	0.0073	0.0098	0.4945	0.0166	0.0075	0.0128	0.8655	0.0759
O_T3	0.0068	0.0108	0.5057	0.0159	0.0073	0.0137	0.8594	0.0706
O_T4	0.0069	0.0098	0.4976	0.0154	0.0077	0.0133	0.8642	0.0705
B_T1	0.0438	0.0431	0.483	0.0187	0.0005	0.0024	0.8731	0.0188
B_T2	0.0438	0.0439	0.4753	0.018	0.0006	0.0026	0.8773	0.018
B_T3	0.046	0.0411	0.471	0.0197	0.0005	0.0023	0.8796	0.0176
B_T4	0.043	0.0436	0.4627	0.0187	0.0005	0.0023	0.884	0.0174

Tao et al.
Tao et al. EURASIP Journal on Advances in Signal Processing 2013 2013:5, doi: 10.1186/1687-6180-2013-5

Analysis of States Separability

The HCA was applied to the learning set, as shown in Figure 13, and the whole process of clustering was displayed in a tree dendrogram. First, each sample was taken as a class. After the first clustering, according to the distance between classes, samples B_1 to B_4 and samples O_1 to O_4 were merged into a cluster, samples I_1 to I_4 were merged into another cluster, and samples N_1 to N_4 were merged into the third cluster. As shown in Table 7, the cluster membership was that all the normal samples gathered in cluster 1, all the inner-race fault samples gathered in cluster 2, all the outer-race fault samples gathered in cluster 3, and all the ball fault samples gathered in cluster 4. The outputs of SPSS Statistics, the tree dendrogram, and the cluster membership table successfully demonstrated the separability of feature vector samples acquired during those four conditions.

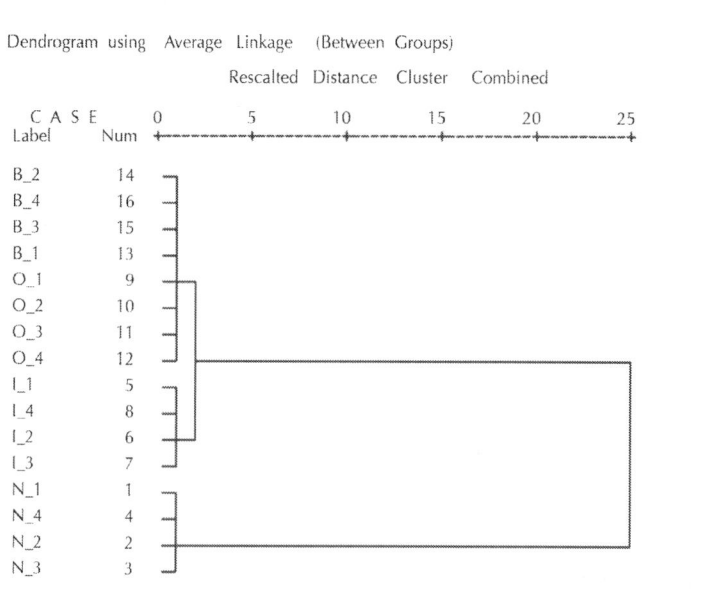

Figure 13: A tree dendrogram of HCA (rolling bearing).

Table 7: Clustering memberships of learning set (rolling bearing)

Case	Cluster index
1:N_1	1
2:N_2	1
3:N_3	1
4:N_4	1
5:I_1	2
6:I_2	2
7:I_3	2
8:I_4	2
9:O_1	3
10:O_2	3
11:O_3	3
12:O_4	3
13:B_1	4
14:B_2	4
15:B_3	4
16:B_4	4

Tao et al.

Tao et al. EURASIP Journal on Advances in Signal Processing 2013 2013:5,

Analysis of Performance Assessment Results

In this study, the normal samples numbered N_1 to N_4 were used to construct and characterize the normal population noted as G_N. Through FDA, the original eight-dimensional space was projected into a new three-dimensional space.

The MDs between the normal population G_N and those samples included in the testing set, as shown in Table 6, were calculated and transformed into normalized CVs according to formula (13). The MD and CV curves are shown in Figures 14 and 15. Apparently, the MDs of the normal testing samples were quite low, while those of the fault testing samples were quite large. Because the normal testing samples were located nearby the normal population G_N, while the inner-race fault, outer-race fault, and ball fault testing samples were located far away. Conversely, the CVs of the normal samples were relatively large, whereas the CVs of all fault testing samples were quite low, falling below the presupposed threshold of 0.6. Thus, it indicated that the inner-race fault, outer-race fault, and ball fault samples were in a faulty condition. Therefore, this result demonstrated that the performance assessment of ball bearing can be quantized and visualized by MD and CV.

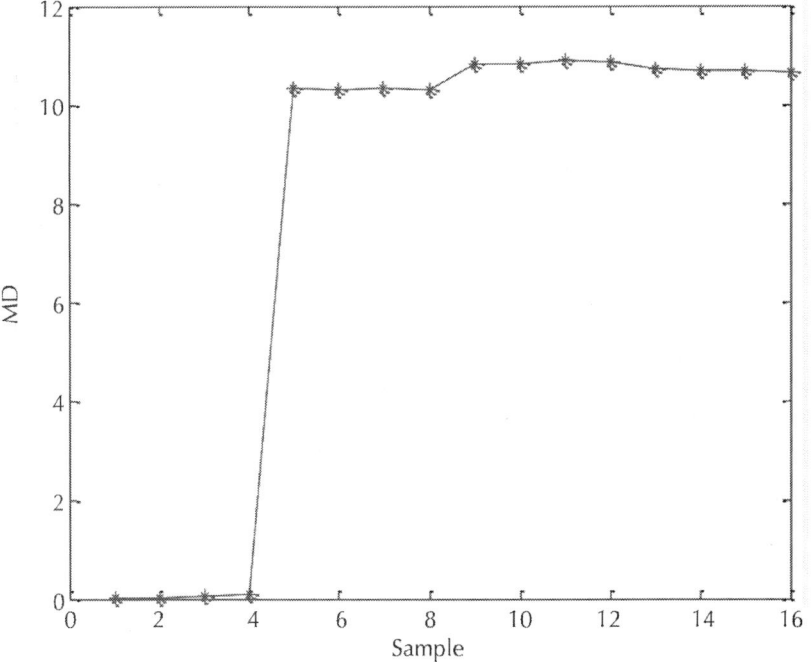

Figure 14: MD result of testing set (rolling bearing).

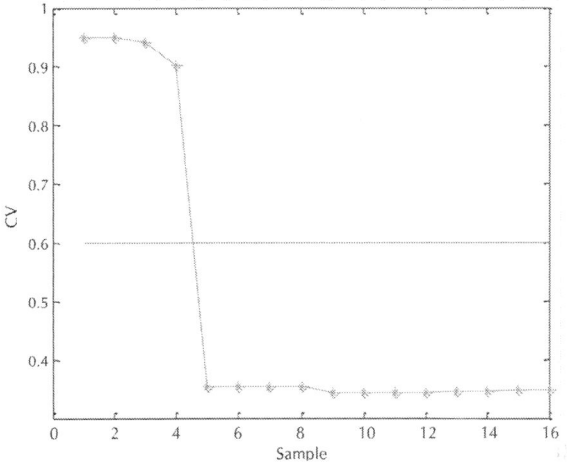

Figure 15: CV result of testing set (rolling bearing).

Analysis of Fault Diagnosis Results

In the performance assessment analysis, the CVs of the samples in the testing set (numbered as I_T1 to I_T4, O_T1 to O_T4, and B_T1 to B_T4) were quite low, so they were in fault states. To identify which type of fault they belonged to, as a reference, the normal samples in the testing set (numbered as N_T1 to N_T4) were also considered, the MDs between the samples in the testing set and those four populations were calculated, as shown in Table 8. In order to facilitate analysis, the normal population was noted as G_N, the inner-race-fault, outer-race-fault, and ball fault populations were noted as G_I, G_O, and G_B, respectively. Through comparative analysis, the samples under normal conditions had the smallest MDs with the population G_N, while the samples under inner-race-fault, outer-race-fault, and ball fault conditions had the smallest MDs with the population G_I, G_O, andG_B, respectively, as shown in the 'Min' row of Table 8. Therefore, it can be determined that the samples N_T1 to N_T4 belonged to normal condition, the samples I_T1 to I_T4, O_T1 to O_T4, and B_T1 to B_T4 were under inner-race fault condition, outer-race fault condition, and ball fault condition, respectively, as shown in the 'Mode' row. This accurate diagnosis result is further proof of the effectiveness of the proposed method in fault diagnosis.

Table 8: MDs and diagnosis results of samples in testing set (rolling bearing)

	N_T1	N_T2	N_T3	N_T4	I_T1	I_T2	I_T3	I_T4
G_N	0.0245	0.0249	0.0352	0.1014	10.3384	10.3041	10.3426	10.2988
G_I	10.3123	10.3094	10.3010	10.3022	0.1079	0.0547	0.0639	0.0609
G_O	10.8671	10.8645	10.8462	10.8709	1.8074	1.9666	1.8590	1.8561
G_B	10.6774	10.6747	10.6590	10.6777	1.2993	1.4575	1.3519	1.3486
Min	0.0245	0.0249	0.0352	0.1014	0.1079	0.0547	0.0639	0.0609
Mode	G_N	G_N	G_N	G_N	G_I	G_I	G_I	G_I
	O_T1	O_T2	O_T3	O_T4	B_T1	B_T2	B_T3	B_T4
G_N	10.8419	10.8391	10.8812	10.8567	10.7244	10.6955	10.6792	10.6498
G_I	1.9145	1.9065	1.9045	1.9098	1.4065	1.3990	1.4010	1.4045
G_O	0.0122	0.0141	0.0358	0.0078	0.5250	0.5308	0.5288	0.5272
G_B	0.5243	0.5168	0.5168	0.5198	0.0642	0.0362	0.0181	0.0150
Min	0.0122	0.0141	0.0358	0.0078	0.0642	0.0362	0.0181	0.0150
Mode	G_O	G_O	G_O	G_O	G_B	G_B	G_B	G_B

Tao et al.
Tao et al. EURASIP Journal on Advances in Signal Processing 2013 **2013**:5, doi: 10.1186/1687-6180-2013-5

CONCLUSIONS

In this article, addressing the challenging issues on performance assessment, fault detection, and fault diagnosis, a novel method based on FDA and MD is introduced, and an integrated framework for performance assessment, fault detection, and fault diagnosis is built.

In this method, FDA is applied as an optimal linear dimensionality reduction technique, in terms of maximizing the separation between different populations. In the new low-dimensional space, MD, which can be transformed into normalized CV, is calculated for performance assessment, and abnormal states can be detected with the presupposed threshold. Furthermore, once various fault data are available, the unknown fault mode can be identified accurately by comparing the MDs between the new data and each normal/fault population.

However, how to transform MD into CV for performance assessment and to determine an adaptive threshold for fault detection is still a challenge for future work. In the future, we are going to acquire sequential online degradation measurements for real-time performance degradation assessment and detection, and enlarge the number of fault and learning samples for more accurate fault diagnosis. Moreover, we plan to apply this approach to different components, such as gearboxes, shafts, etc., to further verify the effectiveness and evaluate the possibility of generalizing the proposed approach.

AUTHORS' CONTRIBUTIONS

XCT carried out the performance assessment studies, and drafted the manuscript; CL (Corresponding author) carried out the diagnosis studies, and participated in the algorithm design and manuscript revision; CL and ZLW carried out the preparation of experimental data, and participated in the algorithm design. All authors read and approved the final manuscript.

ACKNOWLEDGMENTS

This research was supported by the National Natural Science Foundation of China (Grant nos.61074083, 50705005, and 51105019) as well as the Technology Foundation Program of National Defense (Grant no. Z132010B004). The authors are very grateful for the valuable suggestions from the reviewers and editor.

REFERENCES

1. J Lee, J Ni, D Djurdjanovic, H Qiu, HT Liao, Intelligent prognostics tools and e-maintenance. Comput. Ind. 57, 476–489 (2006).
2. AKS Jardine, D Lin, D Banjevic, A review on machinery diagnostics and prognostics implementing condition-based maintenance. Mech. Syst. Signal Process. 20, 1483–1510 (2006).
3. A Heng, S Zhang, ACC Tan, J Mathew, Rotating machinery prognostics: state of the art, challenges and opportunities. Mech. Syst. Signal Process. 23, 724–739 (2009).
4. HP Bloch, FK Geitner, *Practical Machinery Management for Process Plants*, 3rd edn. (Gulf Professional Publishing, Houston, 1997)
5. WB Wang, Modelling the probability assessment of system state using available condition information. IMA J. Manage. Math. 17(3), 225–234 (2006).
6. WB Wang, A two-stage prognosis model in condition based maintenance. Eur. J. Oper. Res.182, 1177–1187 (2007).
7. LL Ma, Z Zhang, JZ Wang, Combination method of support vector machine and fisher discriminant analysis for chemical process fault diagnosis. *Paper presented at the 29th Chinese Control Conference* (Beijing, 2010), pp. 4000–4003
8. LH Chiang, ME Kotanchek, AK Kordon, Fault diagnosis based on Fisher discriminant analysis and support vector machines.

Comput. Chem. Eng. 28, 1389–1401 (2004).
9. QP He, SJ Qin, J Wang, A new fault diagnosis method using fault directions in Fisher discriminant analysis. AICHE J. 51(2), 555–571 (2005).
10. XB He, W Wang, YP Yang, YH Yang, Variable-weighted Fisher discriminant analysis for process fault diagnosis. J. Process. Control. 19, 923–931 (2009).
11. MJ Fuente, D Garcia-Alvarez, GI Sainz-Palmero, Fault detection and identification method based on multivariate statistical techniques. *Paper presented at the Proceedings of Emerging Technologies & Factory Automation* (Mallorca, 2009), pp. 1–6
12. LX Liao, J Lee, Design of a reconfigurable prognostics platform for machine tools. Expert Syst. Appl. 37, 240–252 (2010).
13. I Gurrutxaga, I Albisua, O Arbelaitz, JI Martin, J Muguerza, JM Perez, I Perona, An efficient method to find the best partition in hierarchical clustering based on a new cluster validity index. Pattern Recognit 43, 3364–3373 (2010).
14. J Goldberger, T Tassa, A hierarchical clustering algorithm based on the Hungarian method. Pattern Recognit. Lett. 29(11), 1632–1638 (2008).
15. M Safayani, MTM Shalman, Matrix-variate probabilistic model for canonical correlation analysis. EURASIP J. Adv. Signal Process. 7 (2011) Article ID 748430
16. A Eftekhari, HA Moghaddam, M Forouzanfar, J Alirezaie, Incremental local linear fuzzy classifier in fisher space. EURASIP J. Adv. Signal Process. (2009) Article ID 360834
17. LX Liao, in *An Adaptive Modeling for Robust Prognostics on a Reconfigurable Platform*, ed. by . PhD, University of Cincinnati, Engineering: Industrial Engineering (University of Cincinnati, Cincinnati, 2010)
18. XZ Zhao, BY Ye, Convolution wavelet packet transform and its applications to signal processing. Dig. Signal Process. 20, 1352–1364 (2010).

19. JD Wu, CH Liu, An expert system for fault diagnosis in internal combustion engines using wavelet packet transform and neural network. Expert Syst. Appl. 36, 4278–4286 (2009).
20. YN Pan, J Chen, XL Li, Bearing performance degradation assessment based on lifting wavelet packet decomposition and fuzzy c-means. Mech. Syst. Signal Process. 24, 559–566 (2010).
21. B Kotnik, Z Kacic, A comprehensive noise robust speech parameterization algorithm using wavelet packet decomposition-based denoising and speech feature representation techniques. EURASIP J. Adv. Signal Process. (2007) Article ID 64102
22. G Forestier, C Wemmert, P Gancarski, Multisource images analysis using collaborative clustering. EURASIP J. Adv. Signal Process. (2008) Art ID 374095
23. ZS Wang, A Maier, NK Logothetis, HL Liang, Single-trial classification of bistable perception by integrating empirical mode decomposition, clustering, and support vector machine. EURASIP J. Adv. Signal Process. (2008) Article ID 592742
24. I Kojadinovic, Hierarchical clustering of continuous variables based on the empirical copula process and permutation linkages. Comput. Stat. Data Anal. 54, 90–108 (2010).
25. JAS Almeida, LMS Barbosa, AACC Pais, SJ Formosinho, Improving hierarchical cluster analysis: a new method with outlier detection and automatic clustering. Chemometrics Int. Lab. Syst.87, 208–217 (2007).
26. ML Song, YQ Song, HY Yu, ZY Wang, Calculation of China's environmental efficiency and relevant hierarchical cluster analysis from the perspective of regional differences. Math. Comput. Model., (2012).
27. C Bouveyron, C Brunet, Probabilistic Fisher discriminant analysis: a robust and flexible alternative to Fisher discriminant analysis. Neurocomputing 90, 12–22 (2012).

28. R Khemchandani, Jayadeva, S Chandra, Learning the optimal kernel for Fisher discriminant analysis via second order cone programming. Eur. J. Oper. Res. 203, 692–697 (2010).
29. H Shin, An extension of Fisher's discriminant analysis for stochastic processes. J. Multivariate Anal. 99, 1191–1216 (2008).
30. CR Rao, *Linear Statistical Inference and Its Application*, 2nd edn. (Wiley, New York, 1973)
31. A Heng, A Tan, J Mathew, BS Yang, Machine prognosis with full utilization of truncated lifetime data. *Proceedings of the Second World Congress on Engineering Asset Management*(Harrogate, 2007), pp. 775–784
32. G Niu, S Singh, SW Holland, M Pecht, Health monitoring of electronic products based on Mahalanobis distance and Weibull decision metrics. Microelectron. Reliab. 51, 279–284 (2011).
33. JP Wang, HT Hu, Vibration-based fault diagnosis of pump using fuzzy technique. Measurement39, 176–185 (2006).
34. S Prabhakar, AR Mohanty, AS Sekhar, Application of discrete wavelet transform for detection of ball bearing race faults. Tribol. Int. 35(12), 793–800 (2002).
35. KF Al-Raheem, A Roy, KP Ramachandran, DK Harrison, S Grainger, Rolling element bearing fault diagnosis using Laplace-Wavelet envelope power spectrum. EURASIP J. Adv. Signal Process. (2007) Article ID 073629
36. A Ibrahim, F Bonnardot, M El Badaoui, F Guillet, Detection of bearing damage using stator current, and voltage to cancel electrical noise. EURASIP J. Adv. Signal Process. (2011) Article ID 235236

Chapter 7

A Framework to Determine the Effectiveness of Maintenance Strategies Lean Thinking Approach

Alireza Irajpour[1], Ali Fallahian-Najafabadi[1], Mohammad Ali Mahbod[2], and Mohammad Karimi[3]

[1]Faculty of Management and Accounting, Islamic Azad University, Qazvin Branch, Qazvin, Iran
[2]Department of Management, University of Isfahan, Iran
[3]Islamic Azad University, South Tehran Branch, Tehran, Iran

ABSTRACT

The purpose of this paper is to provide a framework that can identify and evaluate the effectiveness of a given maintenance strategy and to rank components of maintenance system. The framework is developed using DEMATEL method on maintenance strategy as a guideline. To gain a richer understanding of the framework, a questionnaire is constructed and answered by experts. Then the DEMATEL method is applied to analyze the importance of criteria and the casual relations among the criteria are constructed. The scope of the paper is limited to performance measurement of maintenance strategies. It is found that the frame work is applicable and useful for the strategic management of the maintenance function. It is observed that the influencing and preferred infrastructures for designing Learning and Training are three components, that is, optimal maintenance, CMMS, and RCM which are interdependent on each other and are the fundamental components to realize the designed goals of maintenance process. This paper provides an overview of research and developments in the measurement of maintenance performance. Many tools and techniques have been developed in other fields. However, the applicability of those tools to maintenance function has never been tried. In that respect this topic is novel. It helps in managing maintenance more effectively.

INTRODUCTION

Some challenges from modern competitors have provoked many industrial companies to implement new manufacturing approaches [1, 2]. Particularly salient among these is the concept of lean production [3, 4]. Lean production is an approach that includes a set of management practices, including just in time, quality systems, work teams, cellular manufacturing, and supplier management, in an integrated system. The main core of lean production is that these practices can work synergistically to create a high quality system for reaching customer demand with no waste.

Some articles on the topic of lean production system emphasize the relationship between implementation of lean and performance. While most of these studies have focused on a single aspect of lean and its performance implications (e.g., [6–8]), a few studies have explored the implementation and performance relationship with two aspects of lean (e.g., [8, 9]). Even fewer studies have investigated the simultaneous synergistic effects of multiple aspects of lean implementation and performance implication. A noteworthy exception is Cua et al.'s [10] investigation of implementation of practices related to just in time (JIT), total quality management (TQM), and total preventive maintenance (TPM) programs and their impact on operational performance. However, conceptual research continues to stress the importance of empirically examining the effect of multiple dimensions of lean production programs simultaneously [11].

Companies implement lean strategies to achieve better quality, designing the processes which meet customer requirements and expectations, waste elimination (waste is any activity that does not add value to the product or service) and lead time reduction (it helps a Lean enterprise deliver the products to the customer in a shorter time and reduce total costs, both direct and indirect) [12].

Since waste elimination is one of the Lean objectives, it is crucial for companies to identify wastes relevant to defects, waiting time, overproduction (producing more, earlier, or sooner than next workstation demand results in larger inventory and costs), transportation (transportation within Work-In-Process (WIP) resulting from weak plant layout and shortage in understanding of production or process flow), inventory (excess raw materials, finished products, and WIP), unused creativity (failure in exploiting the knowledge and unique abilities of the employees), movement (extra transportation due to wrong location of equipment and tools), and overprocessing (parts of processes that create no added-value to the product or service) [13, 14].

Lean principles have been originated from Toyota's production system known as just in time (JIT) production [15, 16]. The term lean has become widespread after the publication of a book titled The Machine That Changed the World. Then, the term lean production was widely used. Mason-Jones et al. [17] have matched various strategies of supply chain with product type. They have introduced a "leagile" approach which determines the decoupling point between lean and agile paradigms in a supply chain. Sullivan et al. [18] have presented the performance of equipment replacement decision problems within the context of lean manufacturing. They utilized VSM as a road map for providing necessary information for the analysis of equipment replacement decision problem in lean manufacturing implementation.

Muda and Hendry [19] have proposed a world class manufacturing concept incorporated with lean principles for the make-to-order sector. Pavnaskar et al. [20] have presented a classification scheme for lean manufacturing tools. They have suggested that their classifications scheme enables companies to become lean and serve as a foundation for research into lean concepts. Many researchers have contributed to the definition of lean manufacturing. Shah and Ward [11] have provided a comprehensive definition of lean production which is an integrated sociotechnical system whose objective is to eliminate waste by reducing and minimizing the supplier, customer, and internal variability.

The tools and techniques of lean manufacturing include TQM, TPM, Kanban, Kaizen, SMED, Poka-Yoke, and visual control. Houshmand and Jamshidnezhad [21] have presented an extended model of design process of lean production system by means of process variables. They have used axiomatic design theory for developing hierarchical structure to model a design process of lean production system composed of functional requirements, design parameters, and process variables. Braglia et al. [22] have presented a new approach for a complex production system based on seven iterative steps associated with typical industrial engineering tools including VSM. Shah and Ward [23] have defined the measures of lean production. They have mapped the various conceptual lean strategies [24].

Shin et al. [25] have provided the basic data-driven methods including off-line design and on-line computation algorithms; original idea, basic assumption/condition, and computation complexity were presented. Provided methods were implemented on an industrial benchmark.

Evolution of Equipment Management

To begin with, there is a requirement to improve an understanding of the basic perception of the maintenance role. Here, it is pertinent to note that the maintenance function has undergone serious change in the last three decades. The traditional perception of maintenance's role is to fix broken items. Taking such a narrow view, maintenance activities have been confined to the reactive tasks of repair actions or item replacement. Thus, this approach is identified as reactive maintenance, breakdown maintenance, or corrective maintenance. A more recent view of maintenance is defined by Gits [26] as "All activities aimed at keeping an item in, or restoring it to, the physical state considered necessary for the fulfilment of its production function." Clearly, the scope of this opinion also contains the proactive tasks such as the following:

- Routine servicing and periodic inspection,
- Preventive replacement,
- Condition monitoring.

In order to maintain equipment, maintenance must carry out some further activities. These activities contain the planning of work, purchasing and control of materials, personnel management, and quality control [27]. This variety of responsibilities and activities convert maintenance from a simple function to a complex function to manage.

Maintenance should ensure equipment availability in order to produce products at the compulsory quantity and quality levels [28]. The scope of maintenance management includes every phase in the life cycle of technical systems (plant, machinery, equipment, and facilities), specification, acquisition, planning, operation, performance evaluation, improvement, and disposal [29].

Breakdown Maintenance (BM)

This type of maintenance states the maintenance strategy, after the equipment failure equipment is repaired [30]. This maintenance

strategy was mainly implemented in the manufacturing organizations before 1950. In this stage, machines are serviced only when repair is required. This idea has some weaknesses such as the following:
- Unplanned stoppages,
- Excessive damage,
- Spare parts problems,
- High repair costs,
- Excessive waiting and maintenance time,
- High trouble shooting problems [31].

Preventive Maintenance (PM)

This concept is a type of physical checkup of the equipment to prevent equipment breakdown. Preventive maintenance includes activities which are started after a period of time or amount of machine use [32]. This type of maintenance depends on the estimated probability that the equipment will break down in the specified interval. The preventive works are as follows:
- quipment lubrication,
- Cleaning,
- Parts replacement,
- Tightening,
- Adjustment.

Predictive Maintenance (PdM)

Predictive maintenance is often mentioned as condition based maintenance (CBM). In this strategy, maintenance is started in response to a specific equipment condition or performance deterioration [33]. The analytic techniques are organized to measure the physical condition of the equipment such as temperature, noise, vibration, lubrication, and corrosion [34]. When one or more of these indicators reach a set deterioration level, maintenance initiatives are assumed to restore the equipment to desired condition. This means that equipment is taken out of service only when direct evidence exists that deterioration has happened. Predictive maintenance is based on the same principle as preventive maintenance. The advantages of predictive maintenance are based on the need to perform maintenance only when the repair is really necessary, not after a specified period of time [32].

Corrective Maintenance (CM)

The main core of this concept is to prevent equipment failures. This type of maintenance system has been applied to the improvement of equipment; hence the equipment failure can be removed (improving the reliability) and the equipment can be simply maintained (improving equipment maintainability) [35]. The main difference between corrective and preventive maintenance is based on the time of maintenance action. In the corrective action system a problem must exist before corrective actions are taken [36]. The corrective maintenance is following some purposes such as

- Improving equipment reliability,
- Maintainability,
- Safety,
- Reducing design weaknesses (material, shapes),
- Reducing deterioration and failures,
- Aiming at maintenance-free equipment.

Maintenance Prevention (MP)

This type of maintenance system is based on the design phase of equipment. Equipment is designed such that they are maintenance free and an ideal condition of "what the equipment and the line must be" is attained [35]. In the development of new equipment, MP activities must begin at the design stage of equipment [37]. Maintenance prevention often applies some earlier equipment failures and feedback from production areas to ensure equipment design for production systems.

Reliability Centered Maintenance (RCM)

RCM can be defined as an organized, rational process for improving the maintenance requirements of a physical resource in its operating context to understand its "inherent reliability," where "inherent reliability" is the level of reliability which can be attained with an effective maintenance program. RCM is a process implemented to determine the maintenance requirements of any machines or equipment in its operating context by recognizing their functions, the causes of failures, and the effects of the failures.

RCM has seven basic steps:
- Identify the equipment/system to be analyzed;
- Determine its functions;
- Determine what constitutes a failure of those functions;
- Identify the failure modes that cause those functional failures;
- Identify the impacts or effects of those failures' occurrence;
- Use RCM logic to select appropriate maintenance tactics;
- Document your final maintenance program and refine it as you gather operating experience [38].

The various tools employed for affecting maintenance improvement on these 7 steps include
- Failure mode and effect analysis (FMEA),
- Failure mode effect and criticality analysis (FMECA),
- Physical hazard analysis (PHA),
- Fault tree analysis (FTA),
- Optimizing maintenance function (OMF),
- Hazard and operability (HAZOP) analysis.

Productive Maintenance (PrM)

The main aim of productive maintenance is to increase the productivity of a manufacturing unit by decreasing the total cost of the equipment over the whole life from design to equipment degradation. The significant features of this maintenance viewpoint are equipment maintainability and reliability focus, as well as cost reduction of maintenance actions. The maintenance strategy including all previous viewpoints to increase equipment productivity by applying preventive maintenance, corrective maintenance, and maintenance prevention is named productive maintenance [39, 40] (see Figure 3).

Computerized Maintenance Management Systems (CMMSs)

Computerized maintenance management systems (CMMSs) are vigorous for the management of all activities related to the availability, productivity, and maintainability of complex systems. Modern computational facilities have offered a dramatic scope for improved effectiveness and efficiency in, for example, maintenance. Computerized maintenance management systems (CMMSs) have existed, in one form or another, for several decades. CMMS can be used to mechanize the PM function and to help in the control of maintenance inventories and the buying of materials. CMMS can reinforce reporting and analysis capabilities [41, 42]. Accessibility and accuracy of information can provide more reliable decisions in CMMS because of closer working relationships between maintenance and production [43, 44].

Total Productive Maintenance (TPM)

This methodology is linked to the maintenance systems designed and perfected by "Toyota family" companies including Denso and Aisin Seiki [45, 46].

TPM is an innovative approach to maintenance that optimizes equipment effectiveness, removes breakdowns, and promotes autonomous maintenance by operators through day-to-day activities involving total workforce [47]. A strategic approach to improve the performance of maintenance activities is to effectively implement strategic TPM initiatives in the manufacturing organizations.

TPM brings maintenance into attention as a necessary part of the business. The TPM initiative is aimed at improving competitiveness of organizations TPM try to find to involve all levels and functions in an organization to optimize the overall effectiveness of production equipment. This method also tunes up existing processes and equipment by reducing mistakes and accidents. TPM is a world class manufacturing (WCM) initiative that pursues to optimize the effectiveness of manufacturing equipment [48].

Lean TPM and Maintenance Performance Framework

The integration of Lean Thinking [3] and total productive manufacturing (Lean TPM) applies the proven business models of "world class" manufacturing enterprise.

The maintenance performance conceptual framework proposed by Muchiri et al. [49] recognizes main processes that lead the maintenance function to delivery of performance required by manufacturing objectives. The conceptual framework supports alignment of maintenance objectives with the manufacturing and corporate objectives. The conceptual framework has three main sections that include maintenance alignment with manufacturing, maintenance effort/process analysis, and maintenance of results performance analysis.

The first area of the conceptual framework pursues aligning the maintenance objectives with the manufacturing strategy. By studying the requirements of the stakeholders, the performance requirements of the manufacturing system can be well-defined. Cognitive mapping is a crucial tool for studying the cause and effect relationship between strategic essentials [50].

The Maintenance Performance Indicators

The maintenance performance framework summarizes the main essentials that are central in the management of the maintenance function. The essentials make sure that the right work is recognized and effectively implemented for definite results that are in line with the manufacturing performance requirements. Each step is important for effective management of the maintenance function. Both the maintenance process (leading) indicators and maintenance results (lagging) indicators are vital for measuring the performance of the maintenance function. For each essential, the main encounter is to recognize the performance indicators that will express whether the essentials are managed well. Efficient indicators should cover control and monitoring performance and support maintenance actions towards achievement of objectives. Muchiri et al. [49] have provided some indicators that appear often in literature. The classification of Muchiri et al. [49] is applied in this paper for facilitation of performance measurement of maintenance system (see Figure 1).

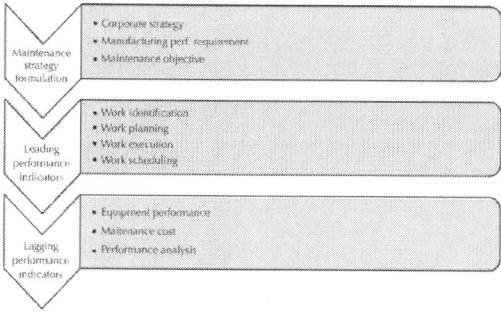

Figure 1: The performance measurement framework for the maintenance function.

Maintenance Process (Leading) Indicators

The maintenance leading indicators monitor some maintenance processes that are as follows:
- Work identification,
- Work planning,

- Work scheduling,
- Work execution.

Key performance indicators for each process are proposed by Muchiri et al. [49] to measure if requirements of each process are satisfied.

Maintenance Results (Lagging) Indicators

The results of the maintenance process can be divided into efficiency of technical systems and cost systems. The lagging indicators are used to measure maintenance results in terms of equipment performance and maintenance cost [49].

DEMATEL Method

DEMATEL method was developed between 1972 and 1979 by the Science and Human Affairs Program of the Battelle Memorial Institute of Geneva, with the purpose of studying the complex and intertwined problematic group.

It has been extensively recognized as one of the methods to explain the cause and effect relationship among the criteria [51–55]. This method is applied to analyze the relationship between cause and effects and among evaluation criteria [56] or to originate interrelationship among factors [55].

Based on Shieh et al. [57], the procedure of DEMATEL method is presented in Figure 2.

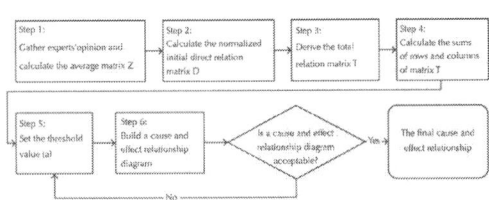

Figure 2: The process of the DEMATEL method (adopted from [5]).

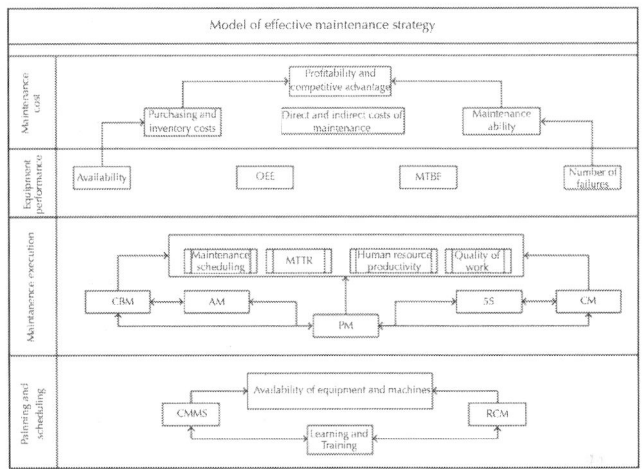

Figure 3: Model of effective maintenance strategy.

RESEARCH METHODOLOGY

This research includes both classes of fundamental and applied researches because it seeks to explain relations and indices and present model. In this research, survey method is used in terms of time and is descriptive (nonexperimental) research in terms of data collection and classification. Case and field research method is used for presentation of model and technique and final conclusion because the present research includes a series of the methods which aims to describe conditions or relations between the studied phenomena on which basis the technique and model are presented by recognizing the previous position and presenting the present position completely.

The present research is conducted with field and library method such as library references including books, publications, theses, and sources available in the university, scientific centers have been used for theoretical information and review of the literature, and direct observation has been used for collecting data.

Research Domain (Theme, Space, and Time)

Business management and plant management problems particularly problems of maintenance management are some of the multidimensional problems and have abundant complexities which may not be evident at the first look. But it becomes evident after deliberation that these problems have deep and extensive dimensions and their multilateral study may be time consuming. Therefore, it is necessary that each researcher first specify limits of his research to prevent confusion and waste of time and sources.

Thematic domain: in this research, researcher tries to study empowering and effective components of maintenance process optimization in manufacturing industries and presents a suitable and practical model for effective application of each component to achieve goals of organization.

Spatial domain: place of this research includes 9 plants, that is, Pol Film and Atlas Film (production of polymer packaging), Plot and Azin Chap (printing and production of heliogravure print cylinder), Zanjan and Bead Wire Industries in the field of wire and welding electrode production, Fardan Aryan Industries (production of food packs and pre-form), MEdisk (production of optical compact disc, CD and DVD) and Arya Kian Industries (automobile steering assembly line).

Time domain: time of the present research is year of 2012 in six-month period and information, statistics, and documents relate to this time period.

Requirement of each applied research is study and recognition of factors affecting working field of research. Information collection instrument is questionnaire. Different maintenance strategies and indices and key elements for implementation of lean maintenance were identified and these concepts were used in two questionnaires.

In questionnaire 1, the experts were asked to specify effect of each lean maintenance element as pairwise comparison and announce in front of each row if these components have been applied in their plants and in what level these components are applied for its implementation. In questionnaire 2, experts were asked to determine effect and significance of each component affecting making maintenance process lean based on four groups of leading indicators and two groups of lagging indicators. This effect is specified according to Saati's scale. It is necessary to note that the following scale is positive for the positive indicators. To determine effect and significance of the component relating to negative indicators, reverse Saati's scale is used.

The indices were determined considering application of interpretive structural modeling approach to ensure theoretical dominance, practical experience, and access due to time-consuming and different types of the questionnaire compared with the common questionnaires. To ensure comprehensiveness of attitudes, the following indices were obtained:

- Relationship between working experience of experts and maintenance issue,
- The presence of experts as maintenance managers and senior experts,
- Experts with the related academic education.

Study of the papers which have used interpretive structural method for analyzing results has suggested the number of experts to be between 4 and 64. 64 selected experts include maintenance managers of 9 plants selected for field studies and 5 experienced maintenance experts are working in these plants.

Data Analysis Method

Extraction of useful results from a research requires application of suitable scientific, accurate, and confirmed methods. In this regard, the following stages are used for analyzing data in this research.

Step 1: It is extraction of approximate agreement matrix for intensity of direct relations between lean maintenance process components adapted from data of questionnaire 6 which was filled by 64 selected experts.

Step 2: This is to structure effect of each lean production component on each other and study feedback and its relations and determine effect of DEMATEL method on lean maintenance process. Application of this method can give the research a suitable structure considering relations between lean maintenance components and make the optimal model policy possible for explaining strategy with lean approach in maintenance process. By extracting two influencing and influenced indices between components of lean maintenance process, effect of components on each other can be evaluated to analyze them properly for ranking these components considering significance of each of them:

- Making direct relations matrix normal,
- Creating general relations matrix,
- Creating cause and effect matrix,
- Creating Dependence matrix,
- Specifying influence order of elements on each other,
- Extracting influencing and influenced indices of lean production components.

Step 3: Approximate agreement matrix is extracted using data of questionnaire 2 and lean maintenance components are regarded as alternative and maintenance leading and lagging indices are regarded as criterion to which output of Step 2, that is, influencing and influenced indices of components, is added.

Step 4: Determining weight of indices using Shannon entropy method: to determine weight of the index, Shannon entropy method is used instead of experts' view due to uncertainty. In this analysis, it specifies weight and significance of each group of leading and lagging indices and influencing and influenced indices of the process lean production components.

Step 5: Ranking and determining weights of lean maintenance components using TOPSIS technique: TOPSIS technique has been used for ranking and weighting because this technique is a compensatory summative method which compares alternatives through weight of each criterion and normalized numbers on each criterion and calculation of the criterion. It is assumed that

TOPSIS has an equal measure for increasing or decreasing values. Normalization usually requires parameters and criteria which almost have heterogeneous dimensions in multicriteria problems. In compensatory methods, such as TOPSIS, exchange between criteria is compulsory so that weak results of a criterion are neutralized and compensated through good results. Compensatory methods preset more realistic form than noncompensatory methods which ignore the solutions obtained through cuts applied on them:
- Creating normalized decision matrix,
- Calculating weighted normal matrix,
- Determining positive and negative ideal point,
- Calculating Euclidean distance of positive and negative ideal solutions,
- Ranking: these numbers are ranked decreasingly to select the preferred option.

Based on output values of TOPSIS, weights of each component are specified using weighted mean method.

Step 6: This step encompasses studying the trend of indices considering implementation levels of lean maintenance components to determine relationship between components of lean maintenances process and trend of maintenances indices in 9 plants selected for field studies, to evaluate effectiveness or loss of efficiency of maintenance process in each one of the plants in six-month period for application or nonapplication of the lean maintenance components.

Model Integration

There is an integration of outputs of DEMATEL and TOPSIS methods that shows elements dependency and ranks that have been specified. Final results of the conducted field studies and analysis of trend between lean maintenances components and trend of maintenances indices have been achieved. An optimized model is presented for explaining maintenance strategy with lean thinking approach and depicted in Figure 3.

RESEARCH FINDINGS AND ANALYSIS

In this way, 9 components have been identified as the factors of which proper implementation can make maintenance process lean and optimal. These components are as follows:
- Inventory management with lean approach,
- 5S,
- CMMS,
- training and learning,
- CBM,
- RCM,
- AM,
- PM,
- CM.

Extraction of approximate agreement matrix for intensity of direct relationship between components of lean maintenance process.

9 × 9 the following pairwise comparison matrix of which components are taken from the studies conducted in review of the literature is a matrix 9 which has been taken from data of questionnaire 6 by 64 selected experts based on group decision making entitled approximate agreement matrix for intensity of direct relationship between components of lean maintenance process. The results have been shown is Table 1.

Table 1: Pairwise matrix of Lean elements of maintenance

Lean elements of maintenance	Inventory management with lean approach	5S	CMMS	Training and learning	CBM	RCM	AM	PM	CM
Inventory management with lean approach	0	3	1	2	2	2	2	4	4
5S	4	0	3	3	3	2	4	3	3
CMMS	4	3	0	2	4	4	2	4	4
Training and learning	2	4	2	0	3	3	4	3	4
CBM	3	2	3	2	0	4	3	3	4
RCM	3	2	2	4	4	0	3	4	4
AM	1	2	2	2	3	3	0	4	4
PM	4	2	3	2	2	2	3	0	4
CM	4	2	2	1	1	2	2	2	0

Determining Effect of Lean Maintenance Components Using DEMATEL

The following results are obtained by implementing DEMATEL technique as shown in Table 2.

Table 2: Determining effect of Lean maintenance components using DEMATEL

Influence matrix T : 9 criteria				
	Row sum (di)	Column sum (ri)	di + ri	di − ri
Inventory management with lean approach	4.33	5.54	9.87	−1.20
5S	5.42	4.38	9.80	1.04
CMMS	5.82	4.03	9.85	1.79
Training and learning	5.42	3.96	9.38	1.47
CBM	5.22	4.68	9.90	0.54
RCM	5.60	4.75	10.35	0.85
AM	4.61	4.98	9.58	−0.37
PM	4.75	5.80	10.54	−1.05
CM	3.57	6.63	10.20	−3.06

In this step, we specify influence order of elements on each other. Elements of column di indicate hierarchy of the influencing elements and order of elements of column ri indicates hierarchy of the influenced elements. (d-j) indicates position of an element and this position will be certainly influencing in case of positivity (d-j) and will be certainly influenced in case of negativity. (d+j) indicates sum of intensity of an element in terms of influencing element and influenced element. Position of an element will be certainly influencing in case of positivity (d-j) and will be certainly influenced in case of negativity. The above analysis shows that CMMS, RCM, education, and culture building are the most effective on their effective execution in the first level compared with other components. Such analysis specifies that CM and PM, spare parts store and purchase will be mostly influenced by application and execution of other components. It is necessary to note that these three components are certainly influenced by other components. Outputs of di will be used as influencing positive index and ri will be used as influenced negative index in the next step for ranking lean maintenance components. The results have been shown in Tables 3 and 4.

Table 3: Ranking based on influenced indices of lean production components

Ranking based on influenced indices of lean production components	
Component	Column sum (ri)
CM	6.63
PM	5.80
Inventory management with lean approach	5.54
AM	4.98
RCM	4.75
CBM	4.68
5S	4.38
CMMS	4.03
Training and learning	3.96

Table 4: Ranking based on influencing indices of lean production components

Ranking based on influencing indices of lean production components	
Component	Row sum (di)
CMMS	5.82
RCM	5.60
Training and learning	5.42
5S	5.42
CBM	5.22
PM	4.75
AM	4.61
Inventory management with lean approach	4.33
CM	3.57

Extraction of Approximate Agreement Matrix for Evaluating Lean Maintenance Components as Maintenance Indices

Approximate agreement matrix is extracted using data of questionnaire 2 and lean maintenance components are regarded as alternative and maintenance leading and lagging indices are regarded as criterion to which output of Step 2, that is, influenced and influencing indices of components, is added. The indices which are included in dark cells are regarded as positive indices and the indices which are in bright cells are regarded as negative indices.

Determining Weight of Indices Using Shannon Entropy Method

Considering the extracted weights, it is observed that weight of each main class of the maintenance leading and lagging indices is shown in Table 5.

Table 5: Determining weight of indices using Shannon entropy method

Type	Group	Weight
Leading indicator	Work identification	12.32%
	Work planning	12.40%
	Work scheduling	13.24%
	Work execution	20.69%
Lagging indicator	Equipment performance	13.22%
	Maintenance cost	28.12%

Considering the analysis in Table 6, we see that 4 components, that is, PM, CMMS, RCM, and spare parts purchase and store, with lean thinking approach are the most important components for moving toward optimum in maintenance. These four components include 52% of the lean weight of the maintenance process.

Table 6: Ranking by TOPSIS

Ranking by TOPSIS		
Lean production components	cli +	Cumulative weight
PM	0.6422	13.8%
CMMS	0.6286	13.5%
RCM	0.5918	12.7%
Inventory management with lean approach	0.5325	11.5%
CM	0.4955	10.7%
Training and learning	0.4758	10.3%
CBM	0.4479	9.6%
5S	0.4172	9.0%
AM	0.4104	8.8%

MODEL PRESENTATION

Outputs of integration methods have been applied in selecting optimal maintenance strategy, optimal model is presented for explaining maintenance with lean thinking approach.

In this model which was presented as bottom-up method, it is observed that the influencing and preferred infrastructures for designing Learning and Training are three components, that is, optimal maintenance, CMMS, and RCM which are interdependent on each other and are the fundamental components to realize the designed goals of maintenance process. In the next step, other components of lean maintenance are given to realize other goals.

CONCLUSIONS

The presented model is an applied model which can be used in different plants and different production lines for optimizing maintenance process. In the performed analyses, it was observed that each one of the lean maintenance components should be valued differently; significance and weight of each of them should be included properly in budgeting for execution. To execute these components for realizing maintenance goals, one should start with the mentioned infrastructures and then apply other components.

In the further researches based on the methodology introduced in the research, one can replace ANP (analytical network process) with TOPSIS or utilize combinatory statistical analysis of Fisher's test and logistic regression test instead of multicriteria decision methods and compare their results with the present research.

REFERENCES

1. R. W. Hall, Attaining Manufacturing Excellence: Just-in-Time, Total Quality, Total People Involvement, Dow Jones-Irwin, Homewood, Ill, USA, 1987.
2. J. R. Meredith and R. McTavish, "Organized manufacturing for superior market performance," Long Range Planning, vol. 25, no. 6, pp. 63–71, 1992.
3. J. P. Womack and D. T. Jones, Lean Thinking: Banish Waste and Create Wealth in Your Corporation, Simon & Schuster, New York, NY, USA, 1996.
4. J. P. Womack, D. T. Jones, and D. Roos, the Machine That Changed the World, Harper Perennial, New York, NY, USA, 1990.
5. D. Sumrit and P. Anuntavoranich, "Using DEMATEL method to analyze the causal relations on technological innovation capability evaluation factors in Thai technology-based firms," International Transaction Journal of Engineering, Management, & Applied Sciences & Technologies, vol. 4, no. 2, 2012.
6. J. R. Hackman and R. Wageman, "Total quality management: empirical, conceptual, and practical issues," Administrative Science Quarterly, vol. 40, no. 2, pp. 309–342, 1995.

7. D. Samson and M. Terziovski, "Relationship between total quality management practices and operational performance," Journal of Operations Management, vol. 17, no. 4, pp. 393–409, 1999.
8. K. E. McKone, R. G. Schroeder, and K. O. Cua, "Impact of total productive maintenance practices on manufacturing performance," Journal of Operations Management, vol. 19, no. 1, pp. 39–58, 2001.
9. B. B. Flynn, S. Sakakibara, and R. G. Schroeder, "Relationship between JIT and TQM: practices and performance," Academy of Management Journal, vol. 38, no. 5, pp. 1325–1360, 1995.
10. K. O. Cua, K. E. McKone, and R. G. Schroeder, "Relationships between implementation of TQM, JIT, and TPM and manufacturing performance," Journal of Operations Management, vol. 19, no. 6, pp. 675–694, 2001.
11. R. Shah and P. T. Ward, "Lean manufacturing: context, practice bundles, and performance," Journal of Operations Management, vol. 21, no. 2, pp. 129–149, 2003.
12. S. Bhasin and P. Burcher, "Lean viewed as a philosophy," Journal of Manufacturing Technology Management, vol. 17, no. 1, pp. 56–72, 2006.
13. J. Drew, B. McCallum, and S. Roggenhofer, Journey to Lean; Making Operational Change Stick, Palgrave Macmillan, New York, NY, USA, 2004.
14. M. Arashpour, M. R. Enaghani, and R. Andersson, "The rationale of lean and TPM," in Proceedings of the International Conference of Engineering and Operations Management (IEOM ‹10›), Dhaka, Bangladesh, 2010.
15. Z. Tang, R. Chen, and J. I. Xuehong, "An innovation process model for identifying manufacturing paradigms," International Journal of Production Research, vol. 43, no. 13, pp. 2725–2742, 2005.
16. F. K. Pil and T. Fujimoto, "Lean and reflective production: the dynamic nature of production models," International Journal

of Production Research, vol. 45, no. 16, pp. 3741–3761, 2007.
17. R. Mason-Jones, B. Naylor, and D. R. Towill, "Lean, agile or league? Matching your supply chain to the marketplace," International Journal of Production Research, vol. 38, no. 17, pp. 4061–4070, 2000.
18. W. G. Sullivan, T. N. McDonald, and E. M. Van Aken, "Equipment replacement decisions and lean manufacturing," Robotics and Computer-Integrated Manufacturing, vol. 18, no. 3-4, pp. 255–265, 2002.
19. S. Muda and L. Hendry, "Proposing a world-class manufacturing concept for the make-to-order sector," International Journal of Production Research, vol. 40, no. 2, pp. 353–373, 2002.
20. S. J. Pavnaskar, J. K. Gershenson, and A. B. Jambekar, "Classification scheme for lean manufacturing tools," International Journal of Production Research, vol. 41, no. 13, pp. 3075–3090, 2003.
21. M. Houshmand and B. Jamshidnezhad, "An extended model of design process of lean production systems by means of process variables," Robotics and Computer: Integrated Manufacturing, vol. 22, no. 1, pp. 1–16, 2006.
22. M. Braglia, G. Carmignani, and F. Zammori, "A new value stream mapping approach for complex production systems," International Journal of Production Research, vol. 44, no. 18-19, pp. 3929–3952, 2006.
23. R. Shah and P. Ward, "Defining and developing measures of lean production," Journal of Operations Management, vol. 25, no. 4, pp. 785–805, 2007.
24. S. Vinodh and S. K. Chintha, "Leanness assessment using multi-grade fuzzy approach," International Journal of Production Research, vol. 49, no. 2, pp. 431–445, 2011.
25. Y. Shin, S. X. Ding, A. Haghani, H. Hao, and P. Zhang, "A comparison study of basic data-driven fault diagnosis and process monitoring methods on the benchmark Tennessee

Eastman process," Journal of Process Control, vol. 22, no. 9, pp. 1567–1581, 2012.

26. C. W. Gits, "Design of maintenance concepts," International Journal of Production Economics, vol. 24, no. 3, pp. 217–226, 1992.

27. V. Priel, Systematic Maintenance Organization, McDonald and Evan, London, UK, 1974.

28. L. M. Pintelon and L. F. Gelders, "Maintenance management decision making," European Journal of Operational Research, vol. 58, no. 3, pp. 301–317, 1992.

29. M. Murray, K. Fletcher, J. Kennedy, P. Kohler, J. Chambers, and T. Ledwidge, "Capability assurance: a generic model of maintenance," in Proceedings of the 2nd International Conference of Maintenance Societies, pp. 1–5, Melbourne, Australia, 1996.

30. T. Wireman, World Class Maintenance Management, Industrial Press, New York, NY, USA, 1990.

31. A. D. Telang, "Preventive maintenance," in Proceedings of the National Conference on Maintenance and Condition Monitoring, K. Vijayakumar, Ed., pp. 160–73, Government Engineering College, Institution of Engineers, Cochin Local Centre, Thissur, India, Febuary 1998.

32. F. Herbaty, Handbook of Maintenance Management: Cost Effective Practices, Noyes Publications, Park Ridge, NJ, USA, 2nd edition, 1990.

33. D. Vanzile and I. Otis, "Measuring and controlling machine performance," in Handbook of Industrial Engineering, G. Salvendy, Ed., John Wiley & Sons, New York, NY, USA, 1992.

34. R. Brook, "Total predictive maintenance cuts plant costs," Plant Engineering, vol. 52, no. 4, pp. 93–95, 1998.

35. H. R. Steinbacher and N. L. Steinbacher, TPM for America, Productivity Press, Portland, Ore, USA, 1993.

36. L. R. Higgins, D. P. Brautigam, and R. K. Mobley, Maintenance Engineering Handbook, McGraw-Hill, New York, NY, USA, 5th edition, 1995.

37. K. Shirose, TPM for Operators, Productivity Press, Portland, Ore, USA, 1992.
38. J. D. Campbell, the Reliability Handbook, Clifford-Elliott, Ontario, Canada, 1999.
39. Y. Wakaru, TPM for Every Operator, Productivity Press, Portland, Ore, USA, 1988.
40. B. Bhadury, Total Productive Maintenance, Allied Publishers, New Delhi, India, 1988.
41. R. Hannan and D. Keyport, "Automating a maintenance work control system," Plant Engineering, vol. 45, no. 6, pp. 108–110, 1991.
42. T. Singer, "Are you using all the features of your CMMS? Following this even-step plan can help uncover new benefits," Plant Engineering, vol. 53, no. 1, p. 32, 1999.
43. R. Dunn and D. Johnson, "Getting started in computerized maintenance management," Plant Engineering, vol. 45, no. 7, pp. 55–58, 1991.
44. A. Labib, "Computerised maintenance management systems," in Complex System Maintenance Handbook, pp. 417–435, Springer, London, UK, 2008.
45. S. Nakajima, Introduction to Total Productive Maintenance (TPM), Productivity Press, Portland, Ore, USA, 1988.
46. D. McCarthy and N. Rich, Lean TPM: A Blueprint for Change, Butterworth-Heinemann, 2004.
47. B. Bhadury, "Management of productivity through TPM," Productivity, vol. 41, no. 2, pp. 240–251, 2000.
48. K. Shirose, TPM Team Guide, Productivity Press, Portland, Ore, USA, 1995.
49. P. Muchiri, L. Pintelon, L. Gelders, and H. Martin, "Development of maintenance function performance measurement framework and indicators," International Journal of Production Economics, vol. 131, no. 1, pp. 295–302, 2011.

50. F. Ackermann, C. Eden, and I. Brown, the Practice of Making Strategy: A Step by Step Guide, SagePublication, London, UK, 2005.
51. Y. J. Chiu, H. C. Chen, G. H. Tzeng, and J. Z. Shyu, "Marketing strategy based on customer behaviour for the LCD-TV," International Journal of Management and Decision Making, vol. 7, no. 2-3, pp. 143–165, 2006.
52. J. J. H. Liou, G. Tzeng, and H. Chang, "Airline safety measurement using a hybrid model," Journal of Air Transport Management, vol. 13, no. 4, pp. 243–249, 2007.
53. G. H. Tzeng, C. H. Chiang, and C. W. Li, "Evaluating intertwined effects in e-learning programs: a novel hybrid MCDM model based on factor analysis and DEMATEL," Expert Systems with Applications, vol. 32, no. 4, pp. 1028–1044, 2007.
54. W. W. Wu and Y. T. Lee, "Developing global managers› competencies using the fuzzy DEMATEL method," Expert Systems with Applications, vol. 32, no. 2, pp. 499–507, 2007.
55. C. Lin and G. Tzeng, "A value-created system of science (technology) park by using DEMATEL," Expert Systems with Applications, vol. 36, no. 6, pp. 9683–9697, 2009.
56. Y. P. Yang, H. M. Shieh, J. D. Leu, and G. H. Tzeng, "A novel hybrid MCDM model combined with DEMATEL and ANP with applications," International Journal Operational Research, vol. 5, no. 3, pp. 160–168.
57. J. I. Shieh, H. H. Wu, and K. K. Huang, "A DEMATEL method in identifying key success factors of hospital service quality," Knowledge-Based Systems, vol. 23, no. 3, pp. 277–282, 2010.

Citations

CHAPTER 1

Rosmaini Ahmad, Shahrul Kamaruddin, Ishak Abdul Azid, Indra Putra Almanar, Failure analysis of machinery component by considering external factors and multiple failure modes – A case study in the processing industry, Engineering Failure Analysis, Volume 25, October 2012, Pages 182-192, ISSN 1350-6307, http://dx.doi.org/10.1016/j.engfailanal.2012.05.007.

CHAPTER 2

Brandon J. Lamarche, Nicole I. Orazio, Matthew D. Weitzman, The MRN complex in double-strand break repair and telomere maintenance, FEBS Letters, Volume 584, Issue 17, 10 September

2010, Pages 3682-3695, ISSN 0014-5793, http://dx.doi.org/10.1016/j.febslet.2010.07.029.

CHAPTER 3

Stanil Y. Ereev and Martin K. Patel, Standardized Cost Estimation for New Technologies (SCENT) - methodology and Tool, ISSN 1613-9623.

CHAPTER 4

Walter J. Weber Jr., Han S. Kim, Removal of greases and lubricating oils from metal parts of machinery processes by subcritical water treatment, The Journal of Supercritical Fluids, Volume 80, August 2013, Pages 30-37, ISSN 0896-8446, http://dx.doi.org/10.1016/j.supflu.2013.03.034.

CHAPTER 5

Kenichi Hitomi, Shigenori Iwai, John A. Tainer, The intricate structural chemistry of base excision repair machinery: Implications for DNA damage recognition, removal, and repair, DNA Repair, Volume 6, Issue 4, 1 April 2007, Pages 410-428, ISSN 1568-7864, http://dx.doi.org/10.1016/j.dnarep.2006.10.004.

CHAPTER 6

Xiaochuang Tao, Chen Lu, Chuan Lu, and Zili Wang, An Approach to Performance Assessment and Fault Diagnosis for Rotating Machinery Equipment, doi:10.1186/1687-6180-2013-5.

CHAPTER 7

Alireza Irajpour, Ali Fallahian-Najafabadi, Mohammad Ali Mahbod, and Mohammad Karimi, "A Framework to Determine the Effectiveness of Maintenance Strategies Lean Thinking Approach," Mathematical Problems in Engineering, vol. 2014, Article ID 132140, 11 pages, 2014. doi:10.1155/2014/132140.

Index

A

Acoustic emissions (AEs) 126
Acrylamide (AM) 2
Acrylic acid (AA) 2, 30
Advanced Land Imager (ALI) 193, 195, 203, 207, 231
Advanced Spaceborne Thermal Emission and Reflection Radiometer (ASTER) 193, 195, 196, 203, 204
Analytical Spectral Devices (ASD) 215

E

Earth Observing One (EO-1) 194
Earth Observing System (EOS) 194, 203
Electromagnetic radiation 172, 173, 183

Enhanced geothermal system (EGS) 119
Enhanced oil recovery (EOR) 2, 30

H

Heavy metal 58, 60, 104

I

Infrared (IR) 2, 5
Institute of Technology Research Institute (IITRI) 183

L

Lithological mapping 194, 195, 196, 197, 212, 213, 215, 216, 219, 223, 226, 227, 228, 282

M

Maximum likelihood classifier (MLC) 217
Mercury 63, 64, 65, 83, 92, 94, 107, 108, 113

N

N-vinyl-2-pyrrolidone (NVP) 2

P

Phytoremediation 59, 65, 66, 76, 89, 90, 96, 99, 100, 101, 102, 105, 106, 107, 108, 109, 110, 111, 112, 113, 114, 282
Phytoremediation technique 65, 99

R

Radio frequency electromagnetic (RF-EM) 178
Remote sensing technology 194
Residual resistance factor (RRF) 30, 34
Resistance factor (RF) 30, 34

S

Scanning electron microscope (SEM) 2, 30, 33
Scanning electron microscopy (SEM) 133
Spectral Angle Mapper (SAM) 212, 213, 219, 220, 221, 224, 225
Spectral Feature Fitting (SFF) 216, 220
Static laser light scattering (SLLS) 41

T

Taupo Volcanic Zone (TVZ) 117, 118
Thermal Infrared Multispectral Scanner (TIMS) 211
Thermochemical sulfate reduction (TSR) 248, 260
Tokyo Sokki Kenkyujo Co. Ltd. (TML) 128
Total depth (TD) 123
Total dissolved solids (TDS) 7
Trillion cubic feet (tcf) 248, 249
True vertical depths (TVD) 123

U

Uniaxial compressive strength (UCS) 128, 129, 141
University of Canterbury (UC) 124

V

Viscosity 172, 181
Visible and near infrared (VNIR) 198, 205, 209

W

Wellbore heating 173